The Institute of Biology's
Studies in Biology no. 43

Introductory Statistics for Biology

by R. E. Parker B.Sc.
Senior Lecturer in Botany,
The Queen's University of Belfast

Edward Arnold

First published 1973
by Edward Arnold (Publishers) Ltd,
25 Hill Street, London W1X 8LL

Boards edition ISBN: 0 7131 2398 2
Paper edition ISBN: 0 7131 2399 0

Printed in Great Britain by
The Camelot Press Ltd, London and Southampton

General Preface to the Series

It is no longer possible for one textbook to cover the whole field of Biology and to remain sufficiently up to date. At the same time students at school, and indeed those in their first year at universities, must be contemporary in their biological outlook and know where the most important developments are taking place.

The Biological Education Committee, set up jointly by the Royal Society and the Institute of Biology, is sponsoring, therefore, the production of a series of booklets dealing with limited biological topics in which recent progress has been most rapid and important.

A feature of the series is that the booklets indicate as clearly as possible the methods that have been employed in elucidating the problems with which they deal. Wherever appropriate there are suggestions for practical work for the student. To ensure that each booklet is kept up to date, comments and questions about the contents may be sent to the author or the Institute.

1973

INSTITUTE OF BIOLOGY,
41 Queen's Gate,
London, S.W.7

Contents

How to Use This Book

Statistics is not presented here as a branch of mathematics but as a logical common-sense development from very simple beginnings. The difficulties, such as they are, do not lie in mathematical manipulation but in grasping a few simple but unfamiliar concepts. Learning to apply statistical methods is rather like learning to swim or to drive a car. One does not become fully proficient immediately and certainly not by just reading and thinking about it. Practical experience is all-important. Through practical experience one acquires practical skills and real understanding.

Do not wait until you have some numerical data to analyse before turning to this book. Start now, at the beginning, work steadily and progressively through it, mastering each new concept as you go, attempting *all* the problems at the ends of the chapters, checking your numerical solutions and logical conclusions with the solutions provided. You will not regret making the effort; the understanding and skill that you acquire will be of lasting value to you both in your biological work and outside.

Belfast 1973 R. E. P.

Probability and Test of Significance 1

1.1 Probability in statistics

In science we often take an everyday word and give to it a rather special meaning, and so it is with 'probability'. In statistics we treat probability quantitatively, having values ranging from zero, which is equivalent to impossibility, to 1.0 which is equivalent to complete certainty. There are two ways of assessing probability:

A priori, when we are dealing with a process which can have one of two or more kinds of outcome and which we understand well enough to be able to predict the probabilities of the different kinds of outcome. For example, we would expect a spun coin to be as likely to fall 'head' as 'tail' side up and that no other result would be feasible. In other words, the probability of a 'head' after a particular spin is equal to the probability of a 'tail' and both equal 0.5. Similarly, in a cross between a heterozygote Aa and a homozygote aa, the expected probability of a single offspring (taken at random) being Aa is equal to the expected probability of it being aa is equal to 0.5. Again, because of the regular cuboid shape of a gambling die we would expect the probability of rolling a 'six' at any one throw to be 1/6 (0.167).

Empirical, when we have no theory on which to predict the probabilities of certain kinds of outcome *or*, if we have, we are interested in testing the theory rather than in assuming it to be true. In such situations we estimate the probability empirically from the observed results of a series of actual trials, as

$$\text{probability } (p) = \frac{\text{number of successes}}{\text{number of trials}}$$

where a trial is a single test such as a single spin of a coin or throw of a die, and a success is the outcome of a trial of the particular type that interests us. The probability estimated is that of achieving a success at any one trial. Estimates made in this way are not precise and are subject to chance variation. In general, estimates tend to become more reliable and to approximate more closely to the *true* value for p as the number of trials increases. It should be remembered that the probability estimated is the probability of an event occurring *before the trial has been made*. After the trial has been made the outcome of that particular trial is known and the probability concept has not the same application. It is of little interest to a man who has just won £100 000 on the football pools to be told that he had only a 0.000 000 1 chance of winning!

1.2 'Frequency' and 'expected' values

We often use statistics to test hypotheses about the way in which systems operate and we do this by comparing the actual results of a series of trials with the results that would be expected on the basis of the hypothesis under test. For this we need the predicted number of successes for a given number of trials. We call this the 'expected' number and estimate it as $p \times N$ where N is the number of trials to be carried out in the test.

Let us suppose that we wish to test whether or not a particular coin spins true. The probability that the coin will fall 'heads' $p_{\text{heads}} = 0.5$, so the expected number of 'heads' will be $0.5 \times N$. If we propose to toss the coin 100 times, the expected number of 'heads' would be 50. Now let us suppose that the coin is tossed and the result is 48 'heads' and 52 'tails'. What would you conclude? Would you say that as the result deviates from the expected 50 'heads' and 50 'tails' the original hypothesis must be rejected, and thus reach the conclusion that the coin did not spin true, *or* would you say that as the result is so near that expected the hypothesis of true spin should be retained? As you will learn later, it is one of the functions of statistics to provide us with a rational basis for decisions of this kind. For the moment you should know that to say that the expected number of 'heads' is 50 is not the same as saying that we expect the number of 'heads' to be 50. There are many possible kinds of outcome from a series of 100 spins, from 50 'heads'/50 'tails' to 100 'heads'/0 'tails' and 0 'heads'/100 'tails'. The probability of the exact 50/50 result is higher than that for any other particular result, but it is still low.

We have seen how we can treat probability in a quantitative manner, how we can estimate it from hypothesis or empirically, or both. We have seen the relationship between the probability of a single event occurring and an expected frequency. We have now to see how we can *add* probabilities, *multiply* them and *partition* them.

1.3 Compound probability

Situations very often exist in which the kind of outcome in which we are interested can come about in a number of different ways. For example, there are a number of different ways of drawing a picture card *of any sort* from a pack of playing cards. In such a case we estimate the total probability by simply adding together the probabilities associated with the individual ways of achieving the result, i.e. $1/52 + 1/52 + 1/52 + 1/52 + \text{etc.} = 12/52 = 3/13$.

This rule is very simple but we have to be a little careful how we apply it. For it to be applicable, the ways in which the stated result can occur must be *exclusive* (i.e. no two can occur together) and *independent* (the occurrence of one must have no effect on the occurrence of another except in the sense implied by exclusiveness).

We often have to deal with situations which lead to one or other kinds of result which are mutually exclusive, e.g. 'head' or 'tail', hit or miss, present or absent, male or female. As there is no other possible result for each situation, the total probability of the two alternative kinds of result is 1.0. Consider a situation in which two men A and B are firing at a single target. Supposing on form we can say that A has a 10% chance of hitting the target with a single shot and B has a 20% chance (i.e. $p_A = 0.1$ and $p_B = 0.2$). What is the probability of the target being hit if both men fire at it? You might say $p_{A+B} = p_A + p_B = 0.3$ or 30%. But before you congratulate yourself, consider a similar situation in which A hits 70% of the time and B hits 80% of the time, i.e. $p_A = 0.7$ and $p_B = 0.8$. On the same basis $p_{A+B} = 1.5$ or 150%. Since a probability of 1.0 is equivalent to certainty, a probability of 1.5 is impossible. What could this result mean?

Before we try to answer this question, let us work out what would happen in the second case, without using probability. Suppose both fire 100 times. We would expect A to hit the target 70 times and to miss 30 times. Out of his 100 shots, we would expect B to hit 80 times and to miss 20 times. If the performance of one man is not influenced by the performance of the other, i.e. the events are independent, we would expect that for the 30 times that A misses, B would miss $20 \times 30/100 = 6$ times. So despite the reasonably good marksmanship of the two men, we would expect the target to be missed by *both* men six times out of the 100. The target would thus be hit 94 times, giving a probability of being hit at any one time of 0.94. (N.B. Using a similar argument for the first set of figures given, the estimated probability of a hit becomes 0.28, as compared with the 0.3 obtained above.) The answers that we have now obtained are the correct ones. This example does not obey the rules for simple addition of probabilities: the events are independent, but *not exclusive*. However poor the marksmanship there is always some chance of the target being hit *twice*, i.e. the two events can occur together. As we shall see later, the total probability of 1.0 associated with a single trial consisting of one shot from both of the two men can be partitioned into three components related to the *three* possible kinds of outcome: a double miss, a single hit and a double hit. The $p = 1.5$ value may be interpreted as indicating that over a long series of trials one could expect the mean number of times the target is hit per volley to approach 1.5.

In contrast to situations in which a particular kind of outcome can come about in a variety of different ways which are independent of one another, we have situations in which the stated kind of result can only come about when a whole series of conditions has been fulfilled. For example, when attempting to open a combination safe we only achieve the desired result when we have satisfied a particular set of conditions, either simultaneously or in succession. Escaping from a maze may present us with a similar situation. In situations of this kind, the probability of the event occurring is estimated as the product of the probabilities of satisfying the conditions individually. Consider the probability of withdrawing, in four successive cards, the four aces from a pack. The individual probabilities are 4/52, 3/51, 2/50 and 1/49, the probability of drawing an ace being reduced as they are removed from the pack. This gives a compound probability of 0.000 037 or 1/27 000. You will see that the application of multiple conditions can greatly decrease the chances of success; failure to realize this soon enough has led to a great many people remaining single!

1.4 Test of significance

It is possible now to indicate the nature of a 'test of significance'. Suppose that a friend were to produce a normal pack of playing cards, declare that the cards had not been marked or tampered with in any way, shuffle them, withdraw four cards in succession apparently at random and then show you that the four cards were the four aces. What would your reaction be? You would have to choose between two alternatives:

(a) that a very unlikely event had just occurred (in fact an event likely to happen in only one in 27 000 trials); or

(b) that a trick has been performed in such a way that the hypothesis of random selection from a normal pack was not applicable. (In other words, you would reject the original hypothesis.)

The choice would be yours. Statistics would simply tell you the probability of the event occurring *if the original hypothesis were true*. With odds as low as 1/27 000, most biologists would conclude they had been tricked.

1.5 Probability distribution

We have already met, in a very simple way, the idea of partitioning the total probability of 1.0 into components (§1.3). When a penny is spun in the air it must come down (probability 1.0) as either a 'head' or a 'tail'. If the coin spins true, the probability of a 'head' is equal to that of a 'tail' at 0.5. Thus the total probability of 1.0 may be partitioned into a probability of 0.5 for 'head' and 0.5 for a 'tail'. As we have seen with two marksmen firing at a target, there are three possible kinds of outcome: no hits, one hit and two hits. Let us now look more carefully and thoroughly into this because this distribution of probability in relation to the different kinds of outcome is the key to a great deal of statistics.

Let us begin with the simplest kind of situation that we can envisage; one in which at a single trial there are only two possible kinds of outcome, e.g. 'head' or 'tail', boy or girl. For our purpose we will assume that the probability of a baby being a girl is equal to that of it being a boy, i.e. for both $p = 0.5$. This fits the *a priori* probability of equal numbers of male and female sex chromosomes on gametogenesis in the male. Consider the following:

Family size	Possible types of family					No. of possibilities
1	M	F				2
	1/2	1/2				
2	MM	MF	FF			3
	1/4	1/2	1/4			
3	MMM	MMF	MFF	FFF		4
	1/8	3/8	3/8	1/8		
4	MMMM	MMMF	MMFF	MFFF	FFFF	5
	1/16	1/4	3/8	1/4	1/16	

You will note that for each family size there is a total probability of 1.0 and that the number of different types of family is always one more than the number in the family. The probabilities of the kinds of family represented in the centre of the table are higher because of the number of different ways of reaching the same result, e.g. a family of two boys and one girl might be acquired as MMF, MFM or FMM. We might represent the case of the family of four by means of a histogram: Fig. 1–1. (N.B. Conventionally bar charts are used to illustrate discontinuous distributions

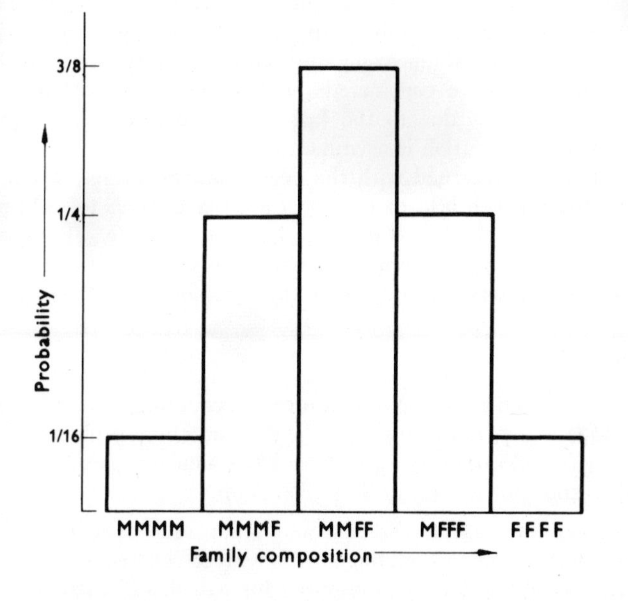

Fig. 1–1 Probabilities of 5 different kinds of family composition for a family of 4 children: $p_{\text{male}} = p_{\text{female}} = 0.5$.

of this kind and histograms reserved for continuous distributions arbitrarily divided into size classes. Since we are concerned here to emphasize similarities rather than differences histograms are used throughout.)

About this we could say:

(a) The probability of having two children of each sex in a family of four is 3/8.

(b) The probability of having *at least one* child of each sex in the family is 7/8 (by summing the three centre terms).

(c) The probability of having all children the same sex is 1/8 (by summing the two outermost terms).

In this simple example we have calculated our probabilities from first principles. In general, in situations of this kind, if we know the number of individuals in a family (n) and the probability of a baby (taken at random) being, say, a boy (p), we can calculate the expected probabilities for all the possible kinds of family by expanding the expression $(p+q)^n$ where $q=1-p$. For $n=4$ this gives:

$$(p+q)^4 = p^4 + 4p^3q + 6p^2q^2 + 4pq^3 + q^4$$

which for $p=0.5$ gives the values in the table above. You may recognize this as the binomial expansion.

The distribution of probability in cases of this kind is essentially *discontinuous* because whatever the family size there are only a limited number of different types of outcome. The larger the family (n) the greater the number of possibilities ($n+1$), and the more nearly will the outline of the corresponding histogram approach a smooth curve. In the example which we have considered the distribution of probability was *symmetrical*. This was due to the fact that $p=0.5=q$. More generally $p \neq 0.5$ and the distribution is asymmetrical.

We have been concerned with the occurrence of discrete events, and with hypotheses which lead us to expect a series of trials to yield results stated in terms of a number of different kinds of outcome. This approach has been extended and developed into a major branch of statistics. For us, at the moment, it serves to introduce the partition of probability.

Problems

1–1 Consider rolling a group of four dice and counting a 'six' as a success. We could score 4, 3, 2, 1 or 0 sixes. The distribution of probability would be given by $(p+q)^n$ where $n=4$ and $p=1/6$. Calculate the distribution of probability and plot the result as a histogram.

1–2 In a certain multiple-choice examination, each question has listed five answers, of which one and one only is correct. There are six questions all of equal weight and 50% is required for passing. If a student selects the answer for each question at random, what is the probability that he will pass the examination?

Normal Distribution 2

2.1 The problem of variation

Many biological surveys and experiments yield results in the form of frequencies with which individuals or observations fall into certain categories. Methods of examining such results have been devised and some of these will be introduced in later chapters (especially Chapters 6 and 7). More generally, results take the form of measurements and it is data in this form that we shall now consider.

Suppose, for example, we are to conduct an experiment to compare the effects of two different treatments on height growth of a particular kind of plant. Aware that the plants with which we have to work are not identical and that the conditions under which we have to grow them are not perfectly uniform, we do not rely on treating only two plants but draw from the plants available to us two random samples each consisting of several similar plants, subject one sample to one set of conditions and the other sample to the other set of conditions. After a suitable time we measure the heights of the plants in each sample and record our results. (For a detailed introduction to experimental method and design see HEATH, 1970.)

The results thus take the form of two series of measurements of the individual plants. How do we deal with such measurements? Without thinking much about it we would probably total the measurements in each series, divide by the number in each and obtain two mean values, one for each treatment. What then do we conclude? Consider the following three examples:

	A		B		C	
	1	*2*	*1*	*2*	*1*	*2*
	25.2	30.1	27.4	23.0	27.6	28.3
	24.9	29.9	30.3	30.3	23.8	33.5
	25.1	29.8	22.7	26.4	22.9	27.2
	25.0	30.0	19.8	36.8	28.2	31.1
	24.8	30.2	24.8	33.5	22.5	29.9
Total (Σx)	125.0	150.0	125.0	150.0	125.0	150.0
Mean value ($\bar{x}$)	25.0	30.0	25.0	30.0	25.0	30.0

In Example A, the individual measurements vary very little from their means and there is clearly no overlap between the ranges of values for the two treatments. With such results we would not hesitate to conclude that the difference of 5.0 between the mean values of the samples was a reliable indication that the two treatments had different effects on plant growth and hence plant height. We would also conclude that the difference of 5.0 was a good estimate of the magnitude of this effect. In example B, we have the same sample totals and means but the individual measurements for each sample are very variable and the ranges of the two samples overlap markedly. Although the two sample means again differ (by 5.0), we would hesitate before concluding that this difference was due to the difference in response to the two treatments. Example C has again the same sample totals and means but shows rather less variation within the samples and less overlap.

In biological work, we are often faced with results like those of examples B and C, and we use statistics to help us in drawing sound conclusions from them. For each experiment of this kind there are two alternative conclusions: (1) that the difference between the sample means is really due to a difference in response to the two treatments and (2) that the difference between the sample means is quite fortuitous, being the result of an uncontrolled variation between individuals.

Whichever conclusion we adopt we might be wrong. If we adopt conclusion (1) *incorrectly* we are said to have made a 'Type 1 error'. If we adopt conclusion (2) *incorrectly* we are said to have made a 'Type 2 error'. The use of statistics does not prevent us sometimes drawing the wrong conclusion but it does provide us with a rational basis for reaching conclusions and enables us to predict how often, on average, we are likely to make errors of Type 1 and Type 2.

2.2 The normal curve

We will first consider the reliability of a single sample mean and then the reliability of an observed difference between two sample means. Quite clearly the reliability of a mean is bound up with the variability of the individual measurements and with the number of them that we have to average. We need then some measure of variability. The measure that we use is related to the distribution pattern of the individual measurements about their mean. Fortunately for us there is a strong tendency for such sets of measurements to show the same kind of distribution, the *normal distribution.*

Suppose we had a large number of measurements (say 500) of the same quantity, e.g. plant height for a certain experimental treatment, we could arbitrarily group our measurements into size classes and then plot a frequency histogram to show how the individual measurements were distributed. Provided that our histogram was based on a large number of measurements, we could use it to indicate the distribution of probability of an individual measurement (taken at random) falling into a particular size class. (N.B. The division of measurements into size classes is automatically achieved by making the measurements to a certain degree of accuracy only.) The histogram produced is likely to be more or less symmetrical and to have an upper outline tending towards a bell-like shape. In some respects it would resemble the histogram for a symmetrical binomial distribution, but would differ from this in two important respects. The range of a binomial distribution is fixed by the number of events recorded in each trial (the number of children per family in our example) but the range of a normal distribution is not limited in this way. The number of classes in the binomial distribution is also fixed and the distribution of probability is essentially discontinuous. We could not work out from the histogram in Chapter 1 (Fig. 1–1) the probability of having $2\frac{1}{2}$ boys and $1\frac{1}{2}$ girls in a family of four. On the other hand, by making our measurements to a high degree of accuracy and making enough of them, we could prepare a histogram of a normal distribution, the enclosing line of which would approach closely to a smooth curve. The theoretical curve towards which many natural frequency distributions tend is called the *normal curve*.

2.3 Mean and standard deviation

Clearly the normal curve cannot be defined in the same terms as the binomial histogram. It may be defined in several ways, but the most useful one is by the equation

$$y = \frac{1}{\sigma\sqrt{2\pi}} \, e^{-[(x-\mu)^2/2\sigma^2]}$$

This is a fearsome expression and you will not be called upon to use it. There are, however, important things to note about it. The variables, x and y, are related through *two* parameters, μ and σ. μ is the *mean*, the point about which the distribution is symmetrical, and σ is the *standard deviation*, a measure of the variability or spread of the measurements about this mean. The important practical point here is that the distribution is completely defined by these two parameters; the other quantities, π and e, are constants. Thus, if we can estimate the standard deviation for a particular set of measurements we have the key to the distribution of probability about their mean. It is useful at this point to remind ourselves about the circumstances in which we use statistics. If we have measurements of the heights of two groups of plants and ask the questions, 'What is the mean height of the plants in each group?' and 'What is the difference between these two means?', we do not require statistics, only simple arithmetic. If, on the other hand, the groups of plants are representative samples which have been experimentally treated and we ask the questions, 'What is the mean height which corresponds with each treatment?' and

'Does the difference between the mean heights of the two groups indicate a real difference between the effects of the two treatments?', then we do.

For each treatment we never have more than a limited number of measurements, never more than a finite sample of the potentially infinite number of possible measurements. We assume, however, that our sample is part of a whole population which shows normal variation. We do not *know* the true value of μ but we estimate it as $\bar{x}$ calculated as $\Sigma x/n$. We do not *know* the true value of σ but estimate it as s where s is calculated as:

$$s = \sqrt{\frac{\Sigma(x-\bar{x})^2}{n-1}}$$

There is no need to be bothered by this formula, it has a common-sense basis. Perhaps the most obvious measure of variability would be the 'mean deviation from the mean', i.e. $\Sigma(x-\bar{x})/n$, but since some of the deviations are negative and some positive a method of converting them all to the same sign is needed. The simplest method is to square the deviations. You might wonder why the divisor used is $(n-1)$ and not n. If we knew the true mean μ then the divisor would in fact be n. But in practice we have to estimate μ as $\bar{x}$, and in so doing we lose what is known as a 'degree of freedom'. If we have n values of x and calculate $\bar{x} = \Sigma x/n$, then clearly we have only $(n-1)$ values of x which are independent of $\bar{x}$. Having determined $\bar{x}$, the nth value of x is also fixed. In practice we rarely evaluate s by means of the above formula, because to do so is unnecessarily laborious and often introduces rounding-off errors. Instead we use the algebraically equivalent expression:

$$\text{Estimated standard deviation } (s) = \sqrt{\frac{\Sigma x^2 - (\Sigma x)^2/n}{n-1}}$$

(N.B. This is the most important formula that you will meet in this book. It should be fully understood and memorized. Σx^2 denotes the sum of the squares of all values of x taken singly and $(\Sigma x)^2$ denotes the square of the sum of all values of x.)

We have seen how a normal distribution may be defined in terms of two parameters, a mean value and a standard deviation, and we understand that like the histogram for a binomial expansion the normal curve which fits a particular set of measurements is our guide to the distribution of probability. We now have to learn to use our estimate of standard deviation in making statements about the reliability of a mean. In much the same way that the probability of obtaining a family of a particular composition was given by the heights of the rectangles in the binomial histogram (Fig. 2–1), so the probability of a single measurement (taken at random) falling between two stated values of x is given by the area of the figure beneath the normal curve and between the two vertical lines corresponding to these values of x (Fig. 2–2).

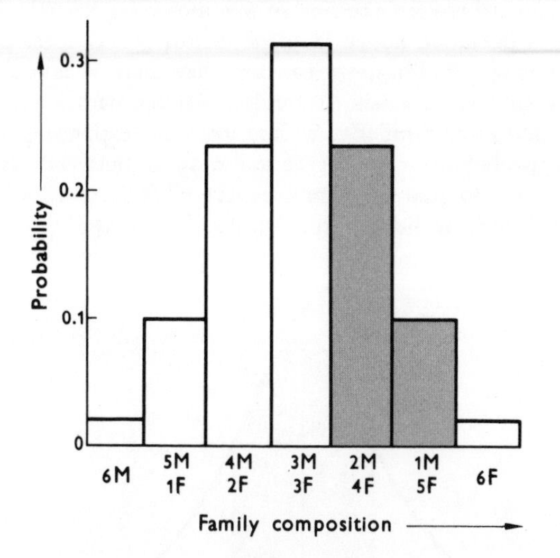

Fig. 2–1 Probabilities of the 7 different kinds of family composition for a family of 6 children: $p_{\text{male}} = p_{\text{female}} = 0.5$. The probability of a family including 1 or 2 boys is indicated by the area of the two shaded rectangles.

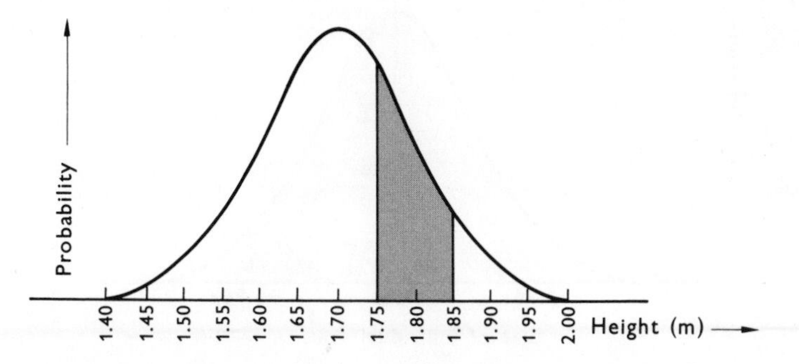

Fig. 2–2 Probability distribution for the heights of human adult males in a population. The probability of a single individual taken at random having a height falling between 1.75 m and 1.85 m is indicated by the area of the shaded part of the figure.

Because the shape of a particular normal curve depends entirely on the standard deviation of the distribution which it represents, it is possible to draw up tables for areas beneath the curve (in terms of probability), between certain limits of x, provided that these limits are expressed in terms of *number of standard deviations* away from the mean. Such tables are then true for *all* normal distributions. For example, we can read from them that the area between the line of symmetry (at $x=\bar{x}$) and the line at one standard deviation above ($\bar{x}+s$) *or* below ($\bar{x}-s$) corresponds to a probability of 0.3413.* Thus, we can say that there is a probability of 0.6826 of a single value taken at random falling within one standard deviation of the mean. Similarly, we can read corresponding figures for $\bar{x}\pm 1.96s$. The probabilities are 0.4750 and 0.95, respectively. In terms of frequency we can say that we would expect 68.26% of all values to lie within $\bar{x}\pm s$ and 95% to lie within $\bar{x}\pm 1.96s$ (Fig. 2–3).

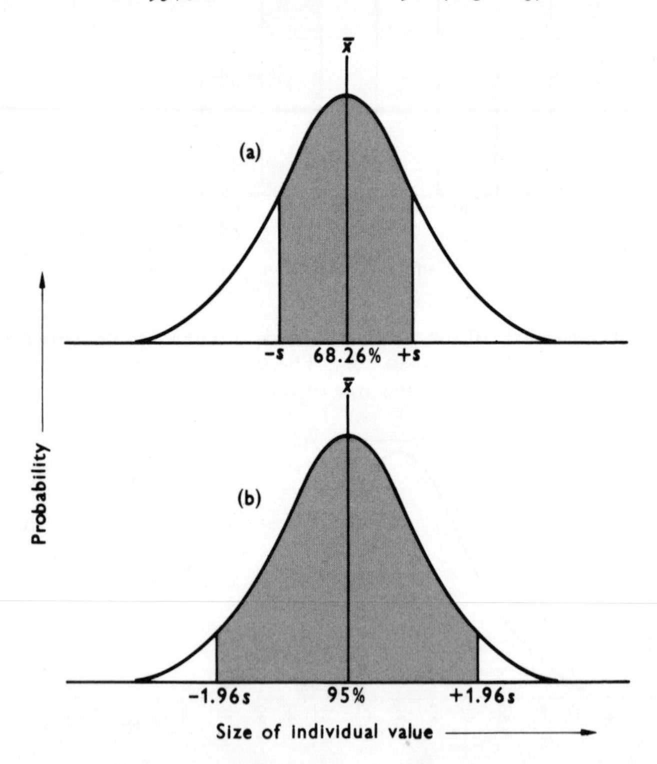

Fig. 2–3 Probability limits for the normal distribution defined in terms of standard deviation (s): **(a)** 68.26% and **(b)** 95%.

Consider for the moment the range $\bar{x}\pm 1.96s$. The probability that an individual value, taken at random, would fall within this range is 0.95; conversely, we can say that given an individual value the probability of the true mean falling within $\pm 1.96s$ of *it* is also 0.95. We thus have a way

* Strictly speaking, this is true only for $\mu\pm\sigma$, but when n is large the estimates $\bar{x}$ and s may be substituted.

of describing the reliability of a single random observation in indicating a mean. We could say that our best (and only) estimate of the mean is the single measurement which we have, and that there is a probability of 0.95 that the true mean lies within the range $\pm 1.96s$ of this.

2.4 Standard error of the mean and confidence limits

It may strike you that to be able to describe the reliability of a single observation in indicating a mean is not a very valuable kind of thing to be able to do. It seems unlikely that we would have an estimate of s unless we had a whole series of measurements from which to calculate it, and that if we had such a series of measurements we would be more interested in the reliability of their mean in indicating the true mean than in the reliability of any one single measurement in doing so. In order to understand how to deal with this we need to know something about the distribution of the means of a number (say n) of measurements. We have seen that if we take a large sample of measurements from a population, divide into a series of size classes and plot a frequency histogram, we would obtain an approximation to a normal curve. If we were now to group these measurements into random groups (of n), calculate the means of these groups and prepare a frequency histogram of these, we would obtain a second histogram which would again approach a normal curve in outline, but with less spread, i.e. with a smaller standard deviation (Fig. 2–4).

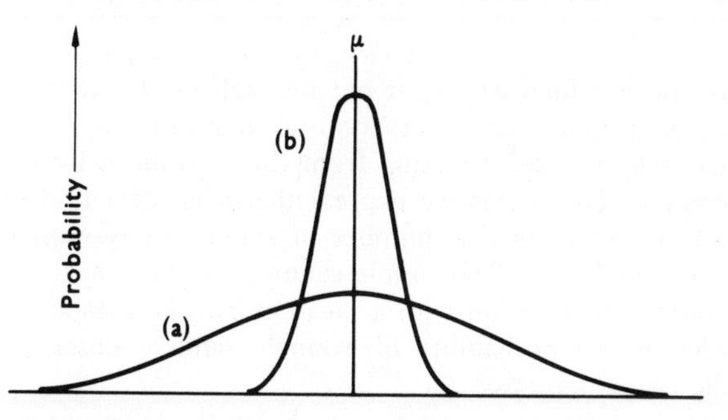

Fig. 2–4 Normal curves: **(a)** for individual values (x), and **(b)** for means ($\bar{x}$) of nine values.

It may be shown that the standard deviation of the distribution of the means of n is equal to $s/\sqrt{n}$. Conventionally the standard deviation of a mean is known as its 'standard error'. In the same way that we can use s in describing the reliability of a single random observation, so we can use $s/\sqrt{n}$ in describing the reliability of a sample mean in indicating the true mean of the whole population. Thus, having calculated $\bar{x}$ we can state that this is our best estimate of the true mean and attach to it *confidence limits* at a chosen level of probability. For example, for the 0.95 level of probability we would calculate $\bar{x} \pm 1.96s/\sqrt{n}$. The true mean then has a probability of 0.95 of falling within these limits and conversely a probability of 0.05 of falling outside.

You will notice that we have selected the probability level of 0.95 for the attachment of confidence limits and have used the multiplier 1.96 for this purpose. Such limits, called '95% confidence limits', are conventional for much biological work. However, we sometimes need to use a different probability level and for this purpose need to select the multiplier which will give us the appropriate partition of probability, e.g. for a probability of 0.99, giving 99% confidence limits, we need to use a multiplier of 2.576 because, for a normal distribution, a total probability of 0.99 lies between $\bar{x} \pm 2.576s$. The general name for these multipliers is 'standardized normal deviate' or d and tables of this quantity are available in which values of d are given for a series of probability levels. Conventionally the probability figures used in these tables are for total probability *outside* the limits. Thus, in the table below the value 1.96 is listed against $p = 0.05$.

Values of d for a series of probability levels p

p	0.1	0.05	0.02	0.01	0.001	0.0001	0.000 01
d	1.645	1.960	2.326	2.576	3.291	3.891	4.417

Another use to which we might put the statistic d is in predicting the probability of an individual observation taken at random, or of a mean of a particular sample size, deviating from the true mean by more than a certain amount. To do this we express the given deviation in terms of standard deviation units (i.e. number of standard deviations) and treat the value obtained as d. Two simple examples should make this clear:

For a normal distribution with a mean of 12.0 and standard deviation of 1.5, what is the probability of a single random observation falling outside the range 9.0–15.0?

Calculate $d = \text{deviation}/s = 3.0/1.5 = 2$. From the table, we see that this value of d corresponds very closely to a probability of 0.05. Thus, there is a probability of approximately 0.05 that a single observation, taken at random, would lie outside the range 9.0–15.0.

For the same distribution, what is the probability of a mean of nine random observations deviating by 2 or more from the mean?

Calculate: $d = \text{deviation}/(s/\sqrt{n}) = 2.0/(1.5/\sqrt{9}) = 2.0/0.5 = 4.0$. From the table, we see that this value exceeds that of 3.891 which corresponds to a probability of 0.0001. Thus, there is a probability of *less than* 0.0001 of a mean of nine random observations deviating from the true mean by 2 or more.

Problems

2–1 Measurements were made on the 30 individuals of a sample and recorded as follows:

12.8	14.1	12.3	15.0	13.7	16.3
11.2	13.5	12.8	11.7	15.9	14.4
13.2	10.3	16.1	14.8	13.7	15.3
13.2	13.9	14.6	12.5	15.3	14.0
13.2	13.8	16.6	12.7	13.4	14.6

Thus $\Sigma x = 414.9$ and $\Sigma x^2 = 5802.47$.

Calculate: (a) the mean ($\bar{x}$), (b) the standard deviation of the individual measurements (s), (c) the standard error of the mean of the 30 measurements and (d) the 95% confidence limits of this mean.

2–2 From the data of Problem 2–1 above, estimate the probability of the sample mean deviating from the true mean by 0.7 or more.

2–3 In a taxonomic study, you wish to estimate the mean length of a particular structure for a population to within ± 1.0 mm with a probability of 95%, You take 36 measurements and calculate the standard deviation from these to be 5 mm. What do you do then?

2–4 Twenty-five determinations of a quantity provide an estimate of the mean with 95% confidence limits of ± 2.0 mg l^{-1}. How many more determinations would be required for the 99% confidence limits to be of this magnitude?

Normal Distribution: continued 3

3.1 Difference between two sample means

The arguments which yielded the method for attachment of confidence limits to single means may be extended to provide a significance test for the difference between two sample means. These developments are perhaps most easily understood in relation to an example.

Suppose that we were studying the variation (or suspected variation) in the performance of a particular plant species with altitude. One of the performance variables which we have studied is 'inflorescence length' and we have measurements for this variable for each of two largish samples: one sample x_1 (with n_1 measurements) from a higher altitude and another x_2 (with n_2 measurements) from a lower altitude. For each sample we could calculate a mean and standard deviation, thus defining the normal distribution corresponding to each sample. Figure 3–1(a) shows two such distributions, one centred on $\bar{x}_1$ and the other on $\bar{x}_2$. You will note that the curves overlap very considerably.

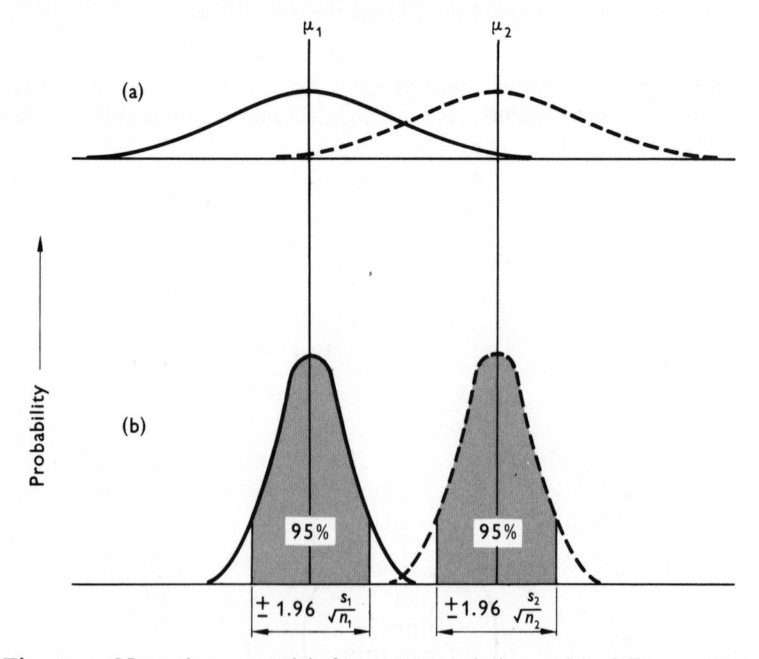

Fig. 3–1 Normal curves: (**a**) for two populations with different means (μ_1 and μ_2) and (**b**) for the means ($\bar{x}_1$ and $\bar{x}_2$) of samples (of n_1 and n_2 values) for the two populations.

Plotted below these (Fig. 3–1(b)) are the distribution curves for the means of n_1 and n_2 measurements respectively. (You will note that these overlap only very slightly.) The central areas beneath these curves to $\pm 1.96s/\sqrt{n}$ of the mean, i.e. those areas including 95% of the total, have been shaded. The figure thus illustrates a situation in which the 95% confidence limits of two sample means do not meet. The means are thus significantly different at a probability greater than 95%.

3.2 Two large samples (n greater than 30)

Whether two sample means are significantly different at a particular level of probability or not may be tested, as the example in Fig. 3–1 suggests, by fitting the appropriate limits to each mean separately and then examining for overlap. This is not an ideal method however. What we really want to know is the probability of obtaining the observed difference between two sample means (or any greater difference) if the two samples are really drawn from normally distributed populations with the same true mean. We have seen in Chapter 2 how we can use the statistic d to estimate the probability of a sample mean deviating from the true population mean by a stated amount (or more); we evaluate

$$d = \frac{\text{deviation of sample mean from true mean}}{\text{estimated standard error of the mean}}$$

and read off the probability corresponding to the value obtained. We can use the same approach to test the significance of a difference between two sample means. When dealing with such a difference, however, we have to take into account the uncertainty associated with both sample means. To do this we treat the difference itself as a normally distributed variable and evaluate:

$$d = \frac{\text{observed difference between sample means}}{\text{estimated standard error of the difference}}$$

The standard error of the difference is clearly greater than the standard error of either mean taken separately (see Fig. 3–1(b)) but it is not simply the sum of the two standard errors. In order to sum measures of variability, we have to use the squared form which is called the *variance*. Thus the variance of x, known as s^2, is given by:

$$s^2 = \frac{\Sigma(x - \bar{x})^2}{n - 1} \equiv \frac{\Sigma x^2 - (\Sigma x)^2/n}{n - 1}$$

Now, the standard deviation of a mean (usually called its standard error) is of the form $s/\sqrt{n}$, so the variance of a mean is $(s/\sqrt{n})^2$ or more simply s^2/n. The variance of the difference between the two sample means

$\bar{x}_1$ and $\bar{x}_2$ is thus (by addition) $s_1^2/n_1 + s_2^2/n_2$ and its standard error is $(s_1^2/n_1 + s_2^2/n_2)^{1/2}$.

So we evaluate:

$$d = \frac{\bar{x}_1 - \bar{x}_2}{\sqrt{[(s_1^2/n_1) + (s_2^2/n_2)]}}$$

and enter the value obtained in the table of d (Table 1) as before.

3·3 Small sample methods

The validity of the procedures just described, i.e. to attach confidence limits to a single sample mean and to test the significance of a difference between two sample means, depends on the assumption that σ, the true standard deviation for the population, may be reliably estimated by s calculated from the sample. When we use probability values taken from the table of the standardized normal deviate (d), we are making this assumption.

When n is large (say, more than 30) we are justified, but when n is small we are not. We need to know the relationship between the probability and deviation/$s\bar{x}$ rather than probability and deviation/$\sigma\bar{x}$. Fortunately this was worked out by William S. Gosset (a statistician from Guinness' Brewery) in 1908. He published under the name of 'Student' and provided us with tables of the required statistic $\boldsymbol{t}$. In simple terms, the probabilities read from the table of $\boldsymbol{t}$ include an allowance for the uncertainty associated with the estimation of the population variance from the sample. For any particular level of probability, $\boldsymbol{t}$ will have a greater value than the corresponding value of d. The value of $\boldsymbol{t}$ corresponding to a particular probability level will depend on the actual sample size, or more precisely on the number of degrees of freedom, being large for small numbers and approaching d as the numbers of degrees of freedom increase.

3·4 Confidence limits to means of small samples

In attaching confidence limits to the means of samples when n is not large, $\boldsymbol{t}$ should be used and *not d*, thus:

$$\bar{x} \pm \boldsymbol{t}\,\frac{s}{\sqrt{n}}$$

where the value of $\boldsymbol{t}$ used is that read for the selected probability level (e.g. 95%) and for $(n-1)$ degrees of freedom.

3·5 Difference between the means of two small samples

When testing the significance of a difference between two sample means, if n_1 or n_2 or both are small (say, less than 30) the ratio of the difference between the two sample means to the standard error of the difference should be equated to $\boldsymbol{t}$ rather than to d (see §3.2). Details of the method vary according to whether we may assume the variances of the two parent populations to be equal or not.

Population variances assumed to be equal. We may often assume that the true variances of the populations of which our two small samples are a part are equal, e.g. when we are dealing with rather small effects of different treatments on samples consisting of organisms drawn originally from the same population. When we may make this assumption we make a combined estimate of this variance using pooled data, thus:

$$s^2 = \frac{1}{n_1 + n_2 - 2}\left(\Sigma x_1{}^2 - \frac{(\Sigma x_1)^2}{n_1} + \Sigma x_2{}^2 - \frac{(\Sigma x_2)^2}{n_2}\right)$$

We then evaluate:

$$\boldsymbol{t} = \frac{\bar{x}_1 - \bar{x}_2}{s\sqrt{[(1/n_1 + 1/n_2)]}}$$

entering the value obtained in the table of $\boldsymbol{t}$ at $(n_1 + n_2 - 2)$ degrees of freedom.

This test will be illustrated by application to the data of example C of §2.1.

Treatments

1	*2*
27.6	28.3
23.8	33.5
22.9	27.2
28.2	31.1
22.5	29.9
$\Sigma x_1 =$ 125.0	$\Sigma x_2 =$ 150.0
$\bar{x}_1 =$ 25.0	$x_2 =$ 30.0
$\Sigma x_1{}^2 = 3154.1$	$\Sigma x_2{}^2 = 4524.2$

$$s^2 = \frac{1}{5+5-2}\left(3154.1 - \frac{125^2}{5} + 4524.2 - \frac{150^2}{5}\right)$$

$$= \frac{1}{8}(3154.1 - 3125 + 4524.2 - 4500)$$

$$= 6.663$$

$$\therefore\ s = 2.581$$

$$\boldsymbol{t} = \frac{5}{2.581\sqrt{\left(\frac{1}{5} + \frac{1}{5}\right)}}$$

$$= \underline{3.063} \text{ with eight degrees of freedom}$$

corresponding to a probability between 0.02 and 0.01.

We conclude that the two means are significantly different, i.e. at a probability of less than 0.05.

Population variances not assumed to be equal. It sometimes happens that we have reason to doubt the equality of the two population variances, either because the samples have been drawn from two independent populations as in many surveys, or because of a large difference between the sample variances. In such cases the pooled estimates of variance, as illustrated above, should not be employed. There is no simple exact solution to the problem thus presented, but an approximate solution is provided by the use of the expression given in §3.2 for large samples, but equating this to t with (n_1+n_2-2) degrees of freedom rather than to d, thus:

$$t = \frac{\bar{x}_1 - \bar{x}_2}{\sqrt{[(s_1{}^2/n_1)+(s^2/n^2{}_2)]}}$$

entering the value obtained in the table of t at (n_1+n_2-2) degrees of freedom.

3.6 Paired-sample test

In the application of the t test which we have considered, the two samples were drawn at random from the populations which they represented. Under certain circumstances the two sets of sample values are related to one another, in that they refer to a single series of individuals or a series of paired individuals. If we wished, for example, to compare the effects of two different feeding regimes on calves, in terms of live-weight gains, we might take a reasonably uniform group of calves, divide it into two samples (or select two samples from it) by means of some randomizing procedure, feed one sample on the first diet and the other sample on the second diet. Our experimental results would be in terms of two series of live-weight gains and we would apply the t test as already explained. (We are unlikely to have enough calves in our samples to justify the application of a simpler d test!) There is, however, a much more efficient method of carrying out the same comparison of diets. We could take a series of calf twins, feed one calf in each pair on the first diet and the other on the second. The results could now be expressed in terms of the *difference* in live-weight gain for each pair of twins. The advantage of the second method is that we would have greatly reduced the component of the variability of our results due to variation in the experimental animals. Similarly, if we wished to compare the effects of two pain-killing drugs on human beings we might arrive at our paired series of results by treating a series of identical twins, or more realistically, by treating the same set of individuals at two different times.

With such paired data we test to see if the mean difference between pairs is significantly different from zero. If we let z be the difference between pairs, we compute:

$$t = \frac{\text{mean difference}}{\text{standard error of the mean difference}} = \frac{\bar{z}}{s_z/\sqrt{n}}$$

where n = number of pairs and enter the value obtained in the table of t at $(n-1)$ degrees of freedom.

(N.B. In computing $\bar{z}$ as $\Sigma z/n$, account is taken of the *sign* of z.)

The value of the paired-sample test may be demonstrated by means of an example.

In an investigation on the effect of eating on pulse rate, a student measured the pulse rate of a group of individuals, ranging from young children to elderly adults, both before and after they had eaten a substantial meal. The following results were obtained, in beats min^{-1}

Subject	1	2	3	4	5	6	7	8	9	10	11	12	13	14
Before	105	79	79	103	87	74	73	82	78	86	77	76	79	104
After	109	87	86	109	100	82	80	90	90	93	81	81	90	110
Diff. (z)	+4	+8	+7	+6	+13	+8	+7	+8	+12	+7	+4	+5	+11	+6

Since both series of measurements relate to the same group of persons, the pair-sample test is the appropriate method of analysis. We test the *mean change* in pulse rate for significance.

$$n = 14; \quad \Sigma z = +106; \quad \Sigma z^2 = 902;$$

$$\bar{z} = +7.571; \quad s_z = \sqrt{\left(\frac{902 - 106^2/14}{13}\right)} = 2.766;$$

$$t = \frac{+7.571}{2.766/\sqrt{14}} = \underline{10.24} \text{ with 13 degrees of freedom}$$

which corresponds to a probability of less than 0.001.

We conclude that the observed *mean increase* in pulse rate with eating is *very highly* significant.

It is instructive to compare this conclusion with that reached by the application of the normal t test. In this case, the following sequence of values is obtained:

$$n_1 = 14; \quad \Sigma x_1 = 1182; \quad \bar{x}_1 = 84.429; \quad \Sigma x_1^2 = 101\,456;$$
$$n_2 = 14; \quad \Sigma x_2 = 1288; \quad \bar{x}_2 = 92.000; \quad \Sigma x_2^2 = 120\,022;$$

$$s^2 = \frac{1}{14+14-2}\left(101\,456 - \frac{1182^2}{14} + 120\,022 - \frac{1288^2}{14}\right) = 122.593$$

$$\therefore \; s = 11.072 \text{ and } t = \frac{7.571}{11.072\sqrt{\left(\frac{1}{14}+\frac{1}{14}\right)}} = \underline{1.81} \text{ with 26 degrees of freedom}$$

which corresponds to a probability more than 0.05.

We conclude that the *increase in mean* pulse rate with eating is ***not*** significant.

Problems

3–1 The measurements (x_1) given in Problem 2–1 are the results for Treatment 1 of an experiment. Treatment 2 yielded another 30 measurements with sum (Σx_2)=388.8 and sum of squares (Σx_2^2)=5096.27. Examine for significance the difference between the two sample means $\bar{x}_1$ and $\bar{x}_2$ by using the statistic *d*.

3–2 Peat soils were analysed for a series of sites classified as having the same type of vegetation. Data for the total phosphate in the peat, expressed in mg/100 g dry weight, are given below for a group of six such sites. Calculate the mean value and attach 95% confidence limits using the statistic ***t***. Total phosphate: 39.3, 46.6, 51.7, 46.0, 68.3, 58.0

3–3 Repeat the test of Problem 3–1 using the statistic ***t***, assuming that the population variances are equal.

3–4 A number of body measurements were made for adult males of a small species of terrestrial mammal, the specimens being taken from two geographically isolated areas (islands) A and B. Only small numbers of suitable animals were available for measurement. The data for tail length, in millimetres, are given below:

A	88, 87, 86, 85, 86, 87, 86, 87, 85, 86, 86, 88, 87, 86
B	87, 85, 85, 86, 84, 85, 84, 86, 83, 85, 86, 83, 85, 84, 87, 85, 84, 86

Examine the significance of the difference between the two sample means using the statistic ***t***, and without assuming equal population variances.

3–5 Complete tests of significance for differences between sample means for examples A and B of §2.1, assuming the population variances to be equal, and given that:

A: $\Sigma x_1^2 = 3125.01$, $\Sigma x_2^2 = 4500.10$ B: $\Sigma x_1^2 = 3191.22$, $\Sigma x_2^2 = 4620.54$

3–6 In studies on *Nardus stricta* (Moor matgrass) it was observed that it grew in well-defined colonies and it was suspected that the grass grew where the soil was *locally deeper*. To test this two soil depth measurements were made, for each of 50 colonies, one within the colony (*Nardus* +) and one just outside (*Nardus* –). The sum of the differences for the 50 pairs of measurements (Σz) = +637 cm. The sum of the squares of these differences (Σz^2) = 14 433. Examine the significance of the mean difference by means of the paired-sample test.

Binomial Distribution 4

4.1 When do we meet this distribution?

We have learnt something about handling data which are normally distributed and before going on to consider data which are distributed differently it would be useful to ask how we have used our knowledge of the normal curve. We have done this by using information from the sample to calculate $\bar{x}$ (sample mean) as an estimate of μ (the population mean), and s (the sample standard deviation) as an estimate of σ (the population standard deviation) which together define a particular normal distribution. The use of s as our measure of variation then allowed us to use tabulated information about the probability distribution common to all normally distributed variables.

We will now consider how to deal in a similar way with data which are binomially distributed. Such data arise when individuals are placed into one or other of two mutually exclusive categories and we make observations on one or more groups of such individuals. You will recall (§1.5) that the probability distribution for such a situation is defined by the terms of the expansion $(p+q)^n$ where p is the probability of a single individual, taken at random, being placed in one particular category, q is the probability of it being placed in the other category, so that $q = 1 - p$, and n is the number of individuals in the group.

4.2 Standard error

Since the distribution is completely determined by p and n, in particular both the *mean* and the *variance* will be so determined. In practice, we will know n and can estimate p from our sample. The mean and probability are related by mean $= p \times n$. It may be shown that the variance of our estimate of p (or q) from a sample is $(p \times q)/n$, and the variance of the number of individuals in groups of n placed in a particular category is $p \times q \times n$. An example will demonstrate the method of use.

Suppose we set up 100 dishes of five seeds (i.e. 500 seeds) in a germination test and obtain the following result:

No. of seeds in dish germinated	5	4	3	2	1	0	Totals
No. of dishes	17	36	31	12	4	0	100
No. of seeds germinated	(85)	(144)	(93)	(24)	(4)	(0)	350

A total of 350 seeds germinate, giving a germination percentage of $350/500 \times 100 = 70\%$. We might reasonably ask how reliable this estimate is. If we assume the seeds to behave independently (i.e. the germination of one seed has no effect on the germination of any other), we can assume the data to be binomially distributed. The variance of our estimate of p (the probability of any one seed germinating) is $(p \times q)/n$, and the standard deviation of our estimate is thus $\sqrt{(pq/n)}$, which in this example $= \sqrt{(0.7 \times 0.3/500)} = 0.0205$. As the germination percentage $= p \times 100$, the standard error of the germination percentage $= 0.0205 \times 100 = \overline{2.05\%}$.*

The usefulness of this kind of approach may be appreciated by comparing the two parallel calculations which follow:

(a) We can base the standard error of the germination percentage on the mean number of seeds germinating per dish and still use the properties of the binomial distribution. The standard deviation of the number of seeds germinating per dish is $\sqrt{pqn}$, where $n = 5$, thus $\sqrt{(0.7 \times 0.3 \times 5)} = \sqrt{1.05} = 1.025$. The standard error of the mean number germinating per dish $= \sqrt{(pqn/N)}$, where N = number of dishes, i.e. $1.025/\sqrt{100} = 0.1025$. As the germination percentage $= 20 \times$ mean number germinating per dish, the standard error of the germination percentage $= 20 \times 0.1025 = \overline{2.05\%}$.

(b) We can also base the standard error on the mean number germinating per dish but calculate it using the 'sum-of-squares' method introduced in Chapter 2. For this we need the results for the individual dishes, thus (cf. solution (b) to Problem 2–1):

Class (c)	Frequency (f)	Class × frequency ($c \times f$)	Class² (c^2)	Class² × frequency ($c^2 \times f$)
5	17	85	25	425
4	36	144	16	576
3	31	93	9	279
2	12	24	4	48
1	4	4	1	4
0	0	0	0	0
	$N = 100$	$\Sigma x = 350$		$\Sigma x^2 = 1332$

Variance of the number of seeds germinating per dish $= \dfrac{1332 - 350^2/100}{99} = 1.081$

* The use of 'standard error' may surprise you. As we assume the distribution to be binomial we can estimate the reliability of a *single determination*, rather than of a *mean* of several determinations, as we did when using the properties of the normal distribution. A formula more convenient for calculation is:

$$\text{standard error of germination percentage} = \sqrt{\left(\frac{10^4 \times p \times q}{n}\right)}$$

Standard deviation of the number of seeds germinating per dish	$=\sqrt{1.081}$	$=1.040$
Standard error of the mean number of seeds germinating per dish	$=1.040/\sqrt{100}$	$=0.104$
Standard error of germination percentage	$=20\times 0.104$	$=\underline{2.08\%}$

4.3 Confidence limits

When dealing with the normal distribution, we were able to attach confidence limits to our sample means using tables of *d* and ***t***. With data which are binomially distributed the attaching of confidence limits raises problems which we must now consider. If we take the expected probabilities of a single dish of five seeds, taken at random, having 5, 4, 3, 2, 1 and 0 seeds germinated by expanding $(0.7+0.3)^5$ and plot these in the form of a histogram (Fig. 4–1), we see that the distribution of probability is not only markedly discontinuous but also far from symmetrical.

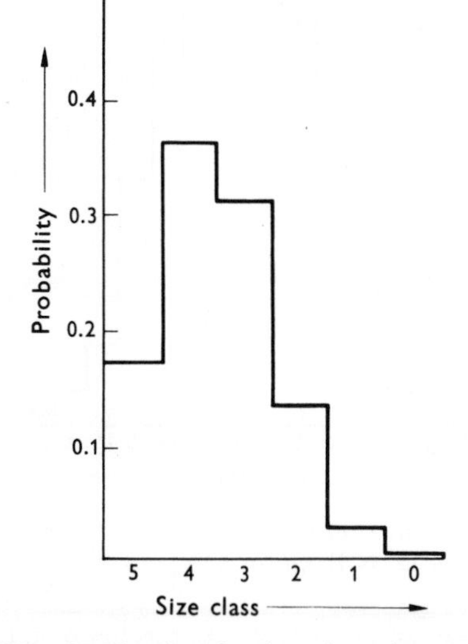

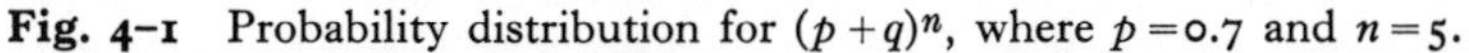
Fig. 4–1 Probability distribution for $(p+q)^n$, where $p=0.7$ and $n=5$.

In placing confidence limits to our estimate of germination percentage, however, we are not dealing with the results as indicated by the behaviour of a group of five seeds but by the behaviour of 100 such groups. It is the probability distribution of results based on groups of 500 seeds that we need.* This is given by the expansion $(0.7+0.3)^{500}$. The histogram for

* The form of the histogram for the distribution of the *mean* number of seeds germinating in 100 dishes each containing five seeds would be exactly the same as that for the distribution of the number of seeds germinating out of 500, but the labelling of the classes would be different.

this distribution would be delimited by a nearly smooth, almost symmetrically curved line. In fact, the binomial distribution tends to normality as n increases, especially if the values of p and q do not closely approach 0 or 1. In practice, then, provided that neither of the products pn (or pnN) or qn (or qnN) is less than 50 we may use the table of the standardized normal deviate, d, to attach confidence limits. The histogram for the expansion $(0.7+0.3)^{30}$ (with $pn=21.0$ and $qn=9.0$) is shown in Fig. 4–2 for comparison with that for $(0.7+0.3)^5$ in Fig. 4–1.

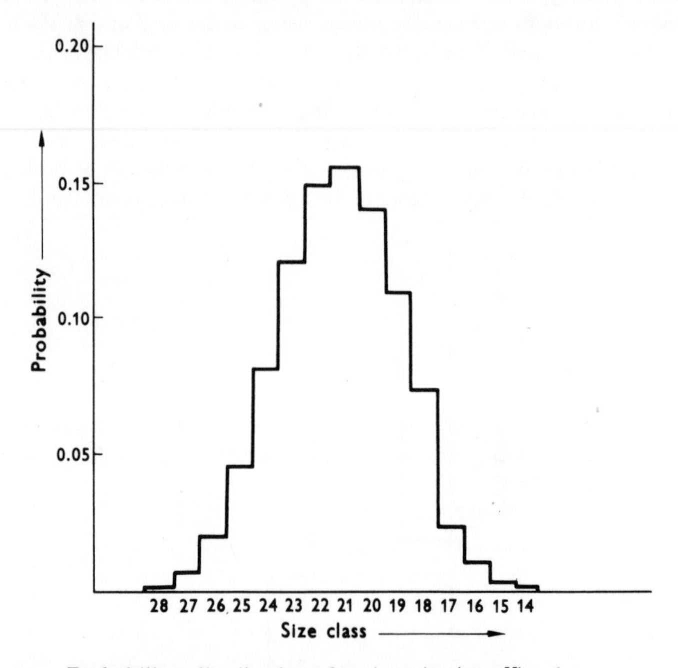

Fig. 4–2 Probability distribution for $(p+q)^n$ (or nN) when $p=0.7$ and n (or nN) $=30$. N.B. qn (or qnN) $=9.0$.

For smaller samples and/or when p approaches 0 or 1, confidence limits may be read directly from Table 41 of PEARSON, E. S. and HARTLEY, H. O. (1962), Table 1.4.1. of SNEDECOR, G. W. and COCHRAN, W. G. (1967), or in Table W of ROHLF, F. J. and SOKAL, R. R. (1969).

4.4 Differences between proportions

In Chapter 3 we learnt how to test the significance of a difference between two sample means for data which were normally distributed; a similar test has been devised for the differences between two proportions or percentages. The method is valid for large samples only as it uses the normal approximation. It takes the same form as the d and t tests in that it involves the evaluation of the ratio (difference between two means)/(standard error of the difference). The difference is between the two *proportions* and the standard error is calculated from pooled data using the fact that the variance of the proportion of each sample is given by pq/n.

Suppose we wish to compare the results of two experimental treatments in breaking dormancy.

	Treatment 1		Treatment 2
Number of seeds set	n_1		n_2
Number of seeds germinated	a_1		a_2
Proportion germinated	$r_1 = a_1/n_1$		$r_2 = a_2/n_2$
Total number of seeds set		$n = n_1 + n_2$	
Overall proportion germinated		$r = \dfrac{a_1 + a_2}{n_1 + n_2}$	

From which we evaluate

$$d = \frac{r_1 - r_2}{\sqrt{[r(1-r)(1/n_1 + 1/n_2)]}}$$

and enter the value obtained in the table of the normal deviate.

Problems

4–1 An experiment was conducted to compare two forms of pre-treatment on the germinability of seeds of a certain species. The treatments were as follows:

(1) Suspended in a *continuous stream* of aerated water for 12 h.

(2) Suspended in a *single volume* of aerated water for 12 h.

A single batch of seed was thoroughly mixed and divided into two more-or-less equal parts. One part was subjected to Treatment 1 and the other part to Treatment 2. After treatment, seeds were sown on moistened seed-testing paper at five seeds per dish, 200 dishes per treatment, and incubated at 20°C for seven days.

The number of seeds germinated in each dish was counted and the results were assembled in the form of two frequency statements, thus:

	Treatment 1						Treatment 2					
No. germinated per dish	5	4	3	2	1	0	5	4	3	2	1	0
Observed frequency (No. of dishes)	30	81	57	24	7	1	9	47	83	57	4	0

Calculations:

(i) For each treatment calculate the total number of seeds germinated $= a$. Then calculate the proportion germinated $(= r)$, the probability of any one seed germinating as a/nN and germination percentage as $100a/nN$, where n = no. of seeds set per dish (= 5), N = no. of dishes set per treatment (= 200) and thus nN = total number of seeds set per treatment (= 1000)

(ii) For each treatment calculate the *expected* probability distribution by expanding:

$$(p+q)^5 = p^5 + 5p^4q + 10p^3q^2 + 10p^2q^3 + 5pq^4 + q^5$$

and convert the probabilities obtained to expected frequencies by multiplying through by N.

Draw histograms for both observed and expected frequencies. Do the observed values fit the expected ones fairly well?

(iii) Assuming that the goodness-of-fit of observed to expected frequencies is acceptable,* calculate the standard errors of the estimates of germination percentage, using the method given at the top of p. 24.

(iv) Using the normal approximation, attach 95% confidence limits to the estimates of germination percentage.

(v) Complete the significance test for the difference between the estimated germination percentages for the two treatments.

4–2 A small boy with 2p in his pocket went to buy some fruit pastilles. He was particularly fond of the blackcurrant ones and had to choose between (a) a single packet of 10 which he knew would include two blackcurrant ones and (b) the same weight of loose pastilles in which the number of blackcurrant ones would be subject to chance.

The shopkeeper told him that if the proportion of blackcurrant ones was the same among the loose ones as in the packets the probability of obtaining less than two blackcurrant pastilles among 10 loose ones taken at random would be greater than of obtaining more than two. Explain.

4–3 In a project to investigate the effect of age on the hatchability of hen eggs, four dozen 4-day-old and the same number of 6-day-old eggs were set in an incubator. A total of 42 of the 4-day eggs hatched and a total of 36 of the 6-day eggs. Is the difference between these two numbers significant?

4–4 The cover percentage of Daisies (*Bellis perennis*) on a particular lawn was estimated by recording 'hits' or 'misses' for 400 randomly-placed pins (= point quadrats). A total of 160 'hits' was recorded. Calculate the mean cover percentage and attach confidence limits to the estimate.

(N.B. Variance of the estimate of $p = pq/n$, and cover percentage $= 100p$.)

4–5 The lawn referred to in Problem 4–4 above was treated with selective herbicide and three weeks later the cover percentage of Daisies was again estimated. On this occasion a total of 120 'hits' was recorded for 400 randomly-placed pins. Calculate the mean cover percentage and test whether the apparent fall in cover percentage is significant.

* We shall learn later how to test goodness-of-fit statistically.

Poisson Distribution 5

5.1 When do we meet this distribution?

There is another distribution which we meet in biology and with which we have to learn to deal. It arises when we make counts of events or individuals which are randomly distributed in space or in time. We most commonly meet it when we are making counts of organisms (or their effects) in area or volume samples, and in counting the emissions of radio-isotopes in 'time samples'. Provided that the items counted nowhere approach maximum possible density, the *frequency* distribution of their counts approaches the discontinuous Poisson series:

$$Ne^{-m}\left(1,\ m\ \frac{m^2}{2!},\ \frac{m^3}{3!},\ \frac{m^4}{4!},\ \ldots \text{ etc.}\right)$$

where N = number of samples examined, m = mean number of items per sample (estimated by $\bar{x}$), e = 2.718 28 and the symbol ! signifies 'factorial'; thus, $3! = 3 \times 2 \times 1 = 6$.

5.2 Standard error

We have seen how a normal distribution may be defined in terms of its mean and standard deviation, and a binomial distribution in terms of its mean (mean = np) and number in a group (n). The Poisson distribution is peculiar in that it is completely defined by one parameter, its mean. Not only this, the variance of the counts is *equal to* the mean. Thus:

Variance of counts = $\bar{x}$
Standard deviation of counts = $\sqrt{\bar{x}}$

Standard error of the mean of N counts = $\dfrac{\sqrt{\bar{x}}}{\sqrt{N}} = \sqrt{\dfrac{\bar{x}}{N}}$

5.3 Confidence limits

The same kind of problem arises over attaching confidence limits as with the binomial distribution and it is solved in a similar way. Poisson distributions with $\bar{x}$ low are very asymmetric and coarsely discontinuous, departing widely from the normal distribution. Histograms for the distributions $\bar{x} = 1.5$ and $\bar{x} = 4.5$ are given in Fig. 5–1.

When placing confidence limits to the mean of N counts, however, it is the distribution of means of N counts which is important. Such distributions are markedly less discontinuous and more nearly symmetrical than

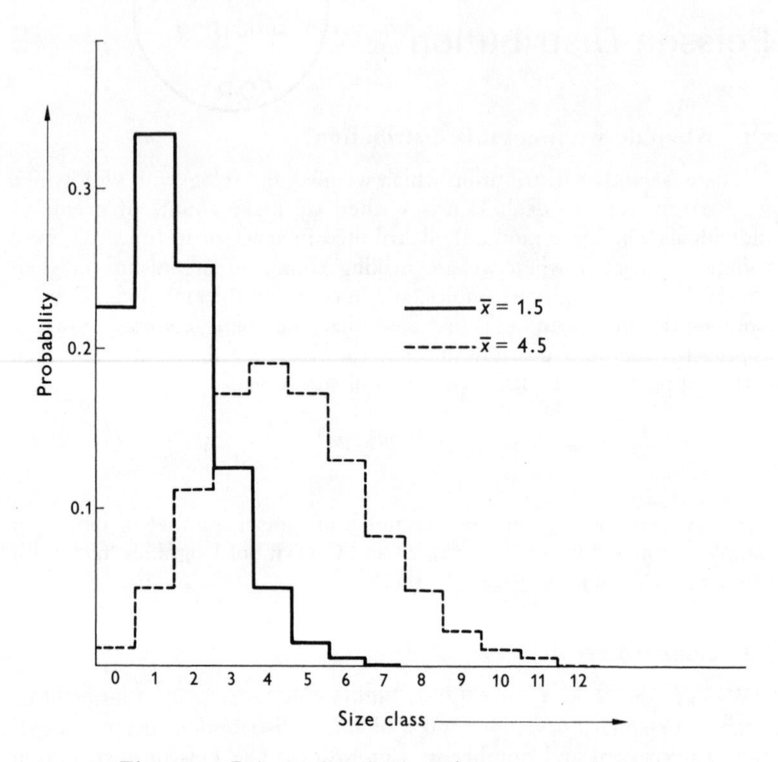

Fig. 5–1 Poisson distributions for $\bar{x} = 1.5$ and $\bar{x} = 4.5$.

those for the individual counts. The histogram for means of 10 counts with $\bar{x} = 1.5$ is given below (Fig. 5–2). (This histogram is the same as for $\bar{x} = 15$ but with the classes differently labelled.)

In practice, we calculate the standard error of the mean from the mean sample count (§5.2) and then, provided that the total number counted ($N\bar{x}$) is large (greater than 50), place confidence limits using the standardized normal deviate, d.

Suppose we wish to estimate the density of algal cells in a certain suspension. After shaking the suspension, we examine samples of it on a haemocytometer slide under a microscope. This permits us to count the algal cells occurring in a series of small samples of equal volume. We may assume the cells to be randomly distributed. Suppose we count the cells in 36 (N) of the 'squares' visible on the slide, and the total number of cells counted to be 198 (Σx). Now:

$$\text{Mean number of cells per square} = \frac{198}{36} = 5{\cdot}5$$

Variance of the 36 counts $= 5{\cdot}5$
Standard deviation of the 36 counts $= \sqrt{5{\cdot}5}$

Standard error of the mean of the 36 counts $= \dfrac{\sqrt{5{\cdot}5}}{\sqrt{36}}$ or $\sqrt{\dfrac{5{\cdot}5}{36}} = \underline{0{\cdot}39}$

Since $N\bar{x}$ is large ($36 \times 5{\cdot}5 = 198$), we use d to attach confidence limits, thus:

$$95\% \text{ confidence limits} = 5{\cdot}5 \pm 1{\cdot}96 \times 0{\cdot}39$$
$$= 5{\cdot}5 \pm 0{\cdot}76, \text{ i.e. } \underline{4{\cdot}74 \text{ to } 6{\cdot}26}$$

Provided that the number of individuals or events counted is large enough, the same procedure may be employed using the count for a single large sample, i.e. when $N = 1$, thus:

Number counted in single sample $= x$
Standard error of the estimated mean number per sample $= \sqrt{x}$
95% confidence limits $= x \pm 1{\cdot}96\sqrt{x}$

When the number counted is small (50 or less), confidence limits may be read directly from Table 40 of PEARSON, E. S. and HARTLEY, H. O. (1962).

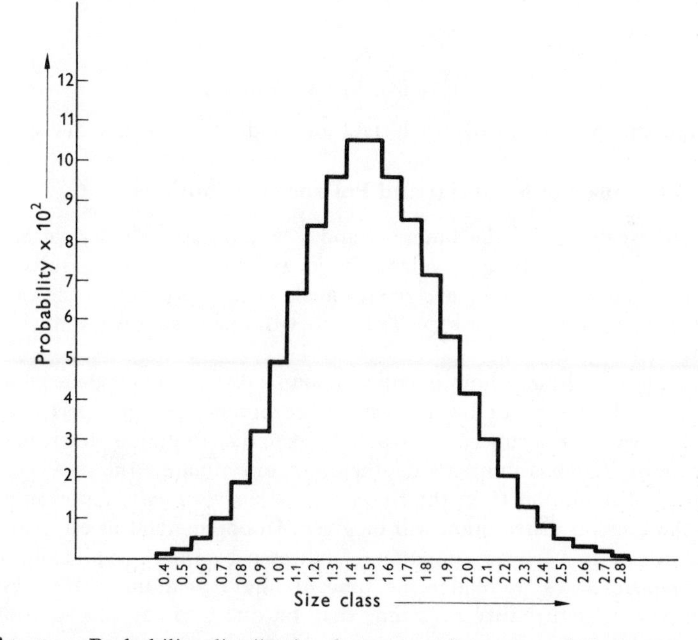

Fig. 5–2 Probability distribution for means of 10 counts (taken at random) from a population with an overall mean of 1.5.

5·4 Difference between two sample means

The significance test for a difference between two means involves the evaluation of the ratio (difference between two means)/(standard error of the difference) as before.

Suppose we wish to compare the mean numbers of bacterial colonies on two sets of agar plates resulting from Treatment 1 and Treatment 2 and the total number of colonies counted is large:

	Treatment 1		Treatment 2
Number of plates examined	N_1		N_2
Total number of colonies counted	Σx_1		Σx_2
Mean number of colonies per plate	$\bar{x}_1 = \Sigma x_1/N_1$		$\bar{x}_2 = \Sigma x_2/N_2$
Difference between two means		$\bar{x}_1 - \bar{x}_2$	
Variance of mean number per plate	$\bar{x}_1/N_1$		$\bar{x}_2/N_2$
Variance of the difference between the two means		$\bar{x}_1/N_1 + \bar{x}_2/N_2$	
Standard error of the difference		$\sqrt{(\bar{x}_1/N_1 + \bar{x}_2/N_2)}$	

We evaluate

$$d = \frac{\bar{x}_1 - \bar{x}_2}{\sqrt{(\bar{x}_1/N_1 + \bar{x}_2/N_2)}}$$

and enter the value obtained into the table of standardized normal deviate, *d*.

5·5 Other uses of binomial and Poisson distributions

In our treatment of the binomial and Poisson distributions, we have concentrated on attaching confidence limits and comparing the results for 'two treatments'. This will have given a misleading impression of the value of these distributions in biology. There are two other main ways in which we use them.

We have already seen how an entire binomial distribution is determined by its probability (p), or by its mean if the number per group is fixed. Similarly, we have seen that the whole Poisson distribution is determined by its mean. The possibility exists, therefore, of estimating the mean (say) for either distribution from the frequency *in one class only*. An example from the Poisson distribution will be given. In one method of estimating mutation rate in bacteria, mutation is indicated by the multiplication of the organism so as to render the tube of liquid medium turbid. The difficulty is that turbidity in a tube may be due to 1, 2, 3, 4 or more mutations. Observation only distinguishes between no mutations on the one hand and one or more mutations on the other. As the mutations occur randomly and the tubes initially contain the same number of bacteria (equal samples), the frequencies with which 0, 1, 2, 3, 4, etc., mutations occur in the tubes will approach those indicated by the terms of a Poisson series. The size of the zero class in such a series is equal to Ne^{-m}, so the mean and hence the total number of mutations may be estimated from the frequency of the zero class only. In practice, of course, the number of bacteria per tube must be adjusted so that the mean number of mutations per tube is very low.

There is another way of using the properties of these distributions which is of greater general importance. In the introductory remarks on each distribution, it was stated that the distributions were only approached by data if certain conditions were satisfied. For the binomial distribution, the individual events recorded must be *independent* of one another. For the Poisson distribution, the events or objects counted must be *randomly distributed*. Significant departure from the expected distribution indicates that these conditions have not been fulfilled. The examination of such departures is an important method of testing and generating hypotheses. A simple example involving the Poisson distribution will be given. If the individual plants of any population are randomly distributed in an area, and the population is sampled by a series of randomly-placed area samples, and the number of plants in each sample counted, the frequency distribution of the counts will approach that of a Poisson series Ne^{-m} $(1, m, m^2/2!$, etc.) where N = number of samples counted and m = the mean count per sample. As we have seen, the variance of such a series is equal to the mean, so if for a series of samples we calculate the mean $\bar{x}$ as $\Sigma x/N$ and the variance s^2 as $[\Sigma x^2 - (\Sigma x)^2/N]/(N-1)$, then we can compare $\bar{x}$ with s^2. If $\bar{x}$ and s^2 thus calculated prove to be more or less equal, we would accept that the plants in the population were randomly distributed. If, however, s^2 proves to be markedly smaller than $\bar{x}$, we would conclude that the plants were more regularly distributed, e.g. because of some self-thinning of the population. If, however, s^2 proves to be markedly greater than $\bar{x}$, we would conclude that the plants tended to be aggregated into patches of relatively high density and we would look for reasons such as limited dispersal of propagules from parents, or concentration of individuals of the species in locally favourable parts of the habitat, to account for this behaviour.

Problems

5–1 The density of a certain bacterial suspension was estimated by means of plate counts. After thorough shaking of the suspension, samples each of 0.5 ml were taken and spread aseptically on the surface of a nutrient agar plate. After incubation, the number of colonies formed on each of 100 plates was counted. The results are given below in the form of a frequency statement:

No. of colonies per plate	0	1	2	3	4	5	6	7	8	9	10	11
Frequency	3	9	17	22	20	14	8	4	2	1	0	0

(i) Calculate the total number of colonies (Σx) and the mean number per plate ($\bar{x}$).

(ii) Calculate the standard error of the mean number per plate, assuming the bacteria to have been randomly distributed in the original inoculum.

(iii) Attach 95% confidence limits to the mean number per plate.

5-2 A baker wishes to make 400 sultana buns. What is the least number of sultanas he may use if the probability that a bun taken at random will have no sultana in it will be less than 0.1%?

5-3 On an average, there are some 1.5 weed seeds in each of a series of vegetable seed packets. What is the probability that a packet taken at random will contain more than three weed seeds?

5-4 In planning to study the seasonal fluctuation in density of the several most important plankton organisms in a pond, a student decided that for each organism he required estimates with 95% confidence limits of about $\pm 5\%$. How many organisms of each species would it be necessary for him to count in each water sample?

5-5 During an autecological study of a particular plant species, counts of the number of individuals present in 250 randomized area samples were made. The results are given below in the form of a frequency statement:

No. of organisms per sample	0	1	2	3	4	5	6	7	8	9	10	11	12
Frequency	43	75	56	34	17	12	9	3	1	0	0	0	0

(i) Calculate the total number of plants counted and the mean number per sample.

(ii) Calculate the expected probability distribution by expanding $e^{-m}(1, m, m^2/2!, m^3/3!$, etc.) using $\bar{x}$ as an estimate of m. Convert the probabilities obtained to expected frequencies by multiplying through by N. Draw histograms for both expected and observed frequencies.
(N.B. $e^{-m} = 1/e^m$, so look up $\log_{10} e$ in tables, multiply this by m to give $m \log_{10} e$. Look up its antilogarithm to obtain e^m, then look up the reciprocal of this to obtain $1/e^m$.)

(iii) Calculate the variance of the sample counts s^2 as $[\Sigma x^2 - (\Sigma x)^2/N]/(N-1)$ and compare s^2 with $\bar{x}$. What do you conclude about the distribution pattern of the plant species?

Use of χ^2: Single Classifications 6

6.1 When do we need χ^2?

The data resulting from biological surveys and experiments are of two main types. In one type, each individual is characterized by a measurement or by some other form of quantitative assessment. In the other type, each individual is placed into a particular class and the results take the form of frequencies with which the individuals fall into these. We have learnt already how to deal with several kinds of measurement data using the statistics *d* and ***t***, and have seen how the methods used may be extended to cover situations in which the assessment of a sample (and hence of its parent population) is made in the form of counts of individuals in certain classes, e.g. as with the estimation of germination percentage. In much biological work, however, the hypotheses which we wish to test do not relate to the quantitative assessment of individuals or samples but directly to the probabilities, and hence the frequencies, of individuals or observations falling into specified classes.

When we wish to test the significance of a deviation of a single ratio between two frequencies from that expected by hypothesis, we may again use *d* (see example of §6.2), but more generally, i.e. when more than two frequencies are involved, we use the statistic χ^2 (chi^2) which was devised by Karl Pearson about 1900. χ^2 is used for a number of different purposes and is evaluated in a variety of ways. It has interesting relationships to the other statistics in common use, but these are disguised in the form in which we usually evaluate it. Its derivation is complex, so we shall concern ourselves primarily with its uses and hope that later it will be possible to understand in retrospect its natural relationships.

6.2 Goodness-of-fit: two classes

Perhaps the simplest kind of situation in which we use χ^2 is the one in which a particular hypothesis leads us to expect individuals or observations to fall into a number of classes with particular frequencies. Observations are made and yield a set of observed frequencies. The problem is then one of testing the significance of any deviation of observed from expected frequencies (i.e. of determining the probability of obtaining such deviation, or any greater, if the hypothesis is true). The form of the statistic which we use is:

$$\chi^2 = \sum \frac{(\mathrm{O} - \mathrm{E})^2}{\mathrm{E}}$$

where O is an observed frequency, and E is the corresponding expected one. The value obtained is entered in the table of χ^2 at $(n - 1)$ degrees of freedom, where n = the number of class frequencies compared, i.e. the number of items in the sum.

We will first consider an example in which two frequencies only are involved. Gregor Mendel raised 929 plants resulting from the self-fertilization of pea plants heterozygous for flower colour; 705 of these had coloured flowers and the remaining 224 had white flowers. Are these numbers of offspring consistent with the hypothesis of a 3:1 ratio of coloured to white?

Class	Observed	Expected	Deviation
Coloured	705	696.75	+8.25
White	224	232.25	−8.25
(Total)	(929)	(929.00)	(0.00)

Using the general formula for χ^2 we have:

$$\chi^2 = \frac{+8.25^2}{696.75} + \frac{-8.25^2}{232.25} = 0.0977 + 0.2931 = \underline{0.3908} \text{ with one degree of freedom}$$

This value corresponds to a probability of rather greater than 0.5, so we conclude that the deviation is *not* significant and the data show an acceptably good fit to hypothesis. (N.B. A deviation as large or larger than the one found would be expected to occur some five or six times out of every 10 trials.)

As the plants fall into two categories only, we could employ the statistic d to make the same test. It then consists of evaluating a ratio of the form (difference)/(standard error of difference) and equating this to d. (cf. Significance test for difference between two sample means: Chapter 3.) As each of the 929 plants must be either coloured or white, the expected frequency distribution is binomial and is given by the expansion $(0.75 + 0.25)^{929}$. (The individual terms giving the probabilities of 0 whites, 1 white, 2 whites, 3 whites, etc., to 929 whites.) The variance of this distribution is given by pqn (§4.2), i.e. $0.75 \times 0.25 \times 929 = 174.19$, and the standard deviation (standard error) for this single result is $\sqrt{174.19} = 13.20$. We then evaluate $d = 8.25/13.20 = \underline{0.625}$ which corresponds, again, to a probability of rather more than 0.5.

(N.B. It is instructive to note in passing that $0.625 = \sqrt{0.3908}$, i.e. d is equal to the square root of χ^2 for one degree of freedom. In the same way that we can equate d = (difference)/(standard error of difference) so we may equate χ^2 = (difference)^2/(variance of difference). For this example,

$$\chi^2 = \frac{8.25^2}{174.19} = \underline{0.3907})$$

6.3 **Correction for continuity** (Yates' correction)

The χ^2 tables to which we refer are calculated from the original distribution, which is continuous. The data for which we use them are essentially discontinuous. There is a resulting tendency to underestimate the final probability and this means that on average we would accept deviations as significant rather more often than they should be (i.e. we should be inclined to make Type 1 errors). This tendency may be corrected by reducing the calculated value of χ^2. We do this by modifying the formula to $\chi^2 = \Sigma(|O-E| - \frac{1}{2})^2/E$, where $|O-E|$ is the positive difference between O and E. The adjustment is only used when there is *one degree of freedom* and should not be used when several χ^2 values are combined.

If we apply the correction to the example of §6.2:

$$\chi^2_{corr} = \frac{(8.25-0.5)^2}{696.75} + \frac{(8.25-0.5)^2}{232.25} = 0.0862 + 0.2586$$

$$= \underline{0.3448} \text{ with one degree of freedom}$$

This value again corresponds to a probability of rather more than 0.5, so the conclusion reached is unaffected by the application of the correction. The correction becomes more important when frequencies are low and deviations are bordering on significance.

6.4 **Goodness-of-fit: more than two classes**

When the observed frequencies are for more than two classes and we wish to test agreement between these observed frequencies and those predicted by hypothesis, the statistic d is not applicable and we are restricted to the use of χ^2. We will consider an example with four classes.

Mendel reported the following results for a dihybrid (double heterozygous) cross with peas: *round/yellow* 315, *round/green* 108, *wrinkled/yellow* 101 and *wrinkled/green* 32 (total 556). Are these results consistent with the hypothesis of independent segregation and the simple dominance of *yellow* over *green* and *round* over *wrinkled*?, i.e. are they consistent with the ratio of 9:3:3:1?

Class	Observed	Expected	Deviation
Round/yellow	315	312.75	+2.25
Round/green	108	104.25	+3.75
Wrinkled/yellow	101	104.25	−3.25
Wrinkled/green	32	34.75	−2.75
(Total)	(556)	(556.00)	(0.00)

Using the general formula for χ^2:

$$\chi^2 = \frac{2.25^2}{312.75} + \frac{3.75^2}{104.25} + \frac{(-3.25)^2}{104.25} + \frac{(-2.75)^2}{34.75}$$

$$= 0.016 + 0.135 + 0.101 + 0.218$$

$$= \underline{0.470} \text{ with three degrees of freedom}$$

This corresponds to a probability of more than 0.9. This means that deviations of this magnitude or greater would be expected to occur more frequently than nine times out of every 10 trials. That such close agreement between observed and expected frequencies should be observed from time to time is to be expected, but if such low values of χ^2, and thus very high probability values, are met with repeatedly bias in recording should be looked for.

6.5 Lower limit of expected frequencies

In the derivation of χ^2, an approximation is involved which holds only when n is 'large'. In practice, it is conventional to use the χ^2 approximation only *when no expected frequency falls below* 5 (but see §7.3). If one or more *expected* frequencies do fall below 5, we pool the smaller groups to form larger ones until the condition is fulfilled. In combining frequency classes in this way we lose degrees of freedom. The number of degrees of freedom for the test is one less than the number of terms in the comparison made after *pooling*. However large the total number of individuals counted, the expected frequencies for the upper terms of a Poisson distribution will always require pooling. For larger values of m, the lowest terms may also have very low expected frequencies (see Fig. 5–2). Pooling is also likely to be required when testing fit to binomial expectation, especially with extreme values of p.

6.6 The effect of fitting a parameter

When the expected frequencies, to which we are examining goodness-of-fit, are determined only by the total number of individuals or observations to be classified and the hypothetical ratios, then the number of degrees of freedom which we have available for making the test is one less than the number of comparisons made. We have considered two such cases involving genetic ratios (§§6.2 and 6.4). It frequently happens, however, that the frequency ratios and therefore the expected frequencies depend on an unknown parameter, e.g. the *mean* (m) in the case of a Poisson distribution and the *probability* (p) in the case of a binomial distribution. These parameters are then estimated from the data themselves, the goodness-of-fit of which is being tested. The effect of estimating a parameter in this way is to reduce by one the number of degrees of freedom for the test. This may be understood by consideration of the following example.

It might be argued that although we are unable to control the sex of our children individually, we might by 'stopping' our families at selected points influence the sex ratio of the population and produce deviations from binomial expectation in family composition. To test this, we might take a large sample of families of one particular size and carry out a goodness-of-fit test between the observed frequencies and those indicated by binomial expectation. There would be two ways of carrying out such a test, rather different hypotheses being tested:

(i) We could test goodness-of-fit to a binomial expansion of $(\frac{1}{2}+\frac{1}{2})^n$ where n would be the number of children in the family size selected. The value of p would be determined by *hypothesis* and the number of degrees of freedom would be *one less* than the number of frequencies compared.* We would be testing *both* the hypothesis that the sex ratio is 1 : 1 *and* that the family composition fits binomial expectation.

(ii) Alternatively, we could estimate the parameter p from the data. We would then be testing *only* that the family composition fits binomial expectation. In this case the number of degrees of freedom available for the test would be *two less* than the number of class frequencies compared.*

(N.B. The family of *one* child is a special case of the above. A type (i) test would test the fit of the sex ratio to the hypothetical 1 : 1, with one degree of freedom. The type (ii) test would have zero degrees of freedom and would be meaningless.)

Problems

6–1 A genetics class examined the *Drosophila* population resulting from the crossing of heterozygous females, w/+ with w/y males, the recessive mutant gene w (white-eyed) being sex-linked.

		♂		
		w	y	
♀	w	$\frac{w}{w}$	$\frac{w}{y}$	(white-eyed)
	+	$\frac{w}{+}$	$\frac{+}{y}$	(red-eyed)
		(♀)	(♂)	

The ratio of red-eyed (wild type) to white-eyed (mutants) was expected to be 1 : 1. The numbers actually counted were: red-eyed 768 and white-eyed 818. Using χ^2, examine the goodness-of-fit of the observed to expected frequencies.

* Assuming no pooling of frequencies.

6–2 In another class, the population of *Drosophila* resulting from crosses of the form, ♀ (vg/ +) × (vg/ +) ♂ (in a heterozygous F_1 population) was examined.

		♂ vg	♂ +
♀	vg	vg/vg	+/vg
♀	+	vg/+	+/+

The mutant gene vg (vestigial-winged) is recessive so the ratio of wild-type (fully-winged) to vestigial-winged was expected to be 3:1. The numbers actually counted were: fully-winged 2492 and vestigial-winged 664. Using χ^2, examine the goodness-of-fit of the observed to expected frequencies.

6–3 Using χ^2, test the goodness-of-fit of the observed to expected frequencies for the seed germination data of Problem 4–1. Remember to amalgamate the classes with the lowest expected frequencies.

If you obtain a value of χ^2 which indicates significant departure of the observed from the expected result, reconsider the details of the experiment and try to suggest a reason for this departure.

6–4 Using χ^2, test the goodness-of-fit of the observed to expected frequencies for the plant occurrence data of Problem 5–4. Remember to amalgamate the classes with the lowest expected frequencies. Does your result confirm that from the (variance)/(mean) ratio test?

Use of χ^2: Double Classifications 7

7.1 Association of attributes: 2 × 2 tables

In our use of χ^2 so far, we have considered cases in which individuals (or observations) have been classified according to one set of criteria only. It frequently happens, however, that we apply two or even more classifications to the same population. As we shall see, χ^2 may again be used to examine certain aspects of the relationships between such simultaneous classifications. When we have individuals classified in several different ways, the question arises as to whether there is any association between one classification and another. We will consider the simplest case, in which there are two classifications each into two mutually exclusive categories.

As part of a study of laterality in Man, the association between handedness and eye-dominance was investigated. Four hundred school children were classified into left-handed or right-handed, and left-eye-dominant or right-eye-dominant. The results were as follows:

		Left-handed	Right-handed	(Total)
Left-eyed	Observed	27	110	(137)
	Expected	18.5	118.5	
Right-eyed	Observed	27	236	(263)
	Expected	35.5	227.5	
(Total)		(54)	(346)	(400)

We may calculate the frequency expected for each of the four classes in the table, using the marginal totals. Thus, a total of 54 children were left-handed and a total of 137 were left-eyed. So, the number of children expected to be *both* left-handed and left-eyed *if the two classifications are independent* is given by $(54 \times 137)/400 = 18.495$ which we round off to 18.5. Once the four expected values have been calculated, we can carry out a goodness-of-fit test between the observed and expected values using the general formula $\chi^2 = \Sigma(O - E)^2/E$. Thus:

$$\chi^2 = \frac{8.5^2}{18.5} + \frac{-8.5^2}{118.8} + \frac{-8.5^2}{35.5} + \frac{8.5^2}{227.5} = \underline{6.866}$$

The value of χ^2 obtained is entered in the table of χ^2 at *one degree of freedom*. It might at first seem that with a four-term comparison we should have three degrees of freedom. In the kind of test we are applying, however, we are using the marginal totals (sums of observed values) to calculate the expected frequencies and this, as with estimating a parameter, leads to a loss in degrees of freedom. Once the marginal totals are fixed, a *single* observed frequency inserted in one cell of the table will determine all of the others. We shall consider later cases in which expected frequencies are calculated from hypothetical ratios instead of marginal totals, and we shall see that in these cases more degrees of freedom remain.

For this example, a χ^2 value of 6.868 with one degree of freedom approximates to the tabulated value of 6.635 which corresponds to a probability of 0.01. Thus we may conclude that there is highly significant association between the two classifications. Since the number of children who are both left-handed and left-eyed exceeds the number expected, we conclude that there is a highly significant association between left-handedness and left-eye-dominance.

In practice, we do not normally use the general formula for computing χ^2 for a 2×2 table. It may be shown that for a 2×2 table:

	A_1	A_2	(Total)
B_1	a	b	$(a+b)$
B_2	c	d	$(c+d)$
(Total)	$(a+c)$	$(b+d)$	$(a+b+c+d)=n$

$$\sum\frac{(O-E)^2}{E}=\frac{n(ad-bc)^2}{(a+c)(b+d)(a+b)(c+d)}$$

With one degree of freedom, Yates' correction for continuity is applicable. The final form of the equation then becomes:

$$\chi^2_{\text{corr.}}=\frac{n(|ad-bc|-\tfrac{1}{2}n)^2}{(a+c)(b+d)(a+b)(c+d)} \text{ with one degree of freedom*}$$

For the above example, we have:

$$\chi^2_{\text{corr.}}=\frac{400(27\times 236-27\times 110-200)^2}{137\times 263\times 54\times 346}$$

$$=\frac{400\times 10\ 252\ 804}{673\ 203\ 204}$$

$$=\underline{6.09}$$

This value falls between the two tabulated values of 6.635 for $p=0.01$ and 5.412 for $p=0.02$. The conclusions are therefore much as before, but the significance level is rather lower.

7.2 Difference between two samples: frequency data

The 2×2 table can be applied to test the significance of a difference between two samples provided they are recorded in terms of numbers of

* $|ad-bc|$ denotes the positive difference between the two products ad and bc.

individuals falling into one or other of two mutually exclusive categories.

Consider the situation presented by Problem 4–5. The cover of Daisies in a lawn was examined both before and after the application of selective herbicide by means of 400 randomly-distributed point quadrats. Totals of 160 and 120 'hits' respectively were recorded. We wish to know whether the apparent fall in cover percentage is significant.

Set up the 2×2 table:

	1st estimate	2nd estimate	(Total)
'Hits'	160	120	(280)
'Misses'	240	280	(520)
(Total)	(400)	(400)	(800)

$$\chi^2_{\text{corr.}} = \frac{800(280 \times 160 - 240 \times 120 - 400)^2}{280 \times 520 \times 400 \times 400}$$

$$= \underline{8.357} \text{ with one degree of freedom}$$

This corresponds to a probability of between 0.01 and 0.001, confirming that the recorded fall in cover is highly significant.

7.3 c × r tables

Tests of association are not restricted to cases in which classifications are into two categories only. The 2×2 table is, in fact, only a special case of the more general c (columns) × r (rows) table. The derived formula for χ^2 as used for the 2×2 table is not applicable, so we normally compute χ^2 in the form $\Sigma(O - E)^2/E$. The number of degrees of freedom available for testing association is $(c - 1)(r - 1)$, and since this exceeds unity for all tables larger than 2×2 the Yates' correction is not applied. The interpretation of association demonstrated by means of a 2×2 table is often quite straightforward because it is either simply positive or simply negative. Significant association in c × r tables, however, is frequently more complex and difficult to interpret (see §8.2), and it is an advantage to have available for inspection the contribution of each cell in the table to the total χ^2 (i.e. its $(O - E)^2/E$ value) so that it is possible to tell at a glance the cells responsible for the major deviations from expectation. If this facility is not required, e.g. if absence of association is confidently expected, another form of χ^2 may be used, i.e. $\chi^2 = \Sigma(O^2/E) - n$.

For the larger tables, the objection to very low expected frequencies (i.e. less than 5) decreases, as the likelihood of an inflated contribution from an individual cell of the table influencing the final total χ^2 is reduced. Provided that expected frequencies do not fall below unity and that no more than a small proportion are less than 5, the conclusions drawn from the analysis are likely to be reliable.

In a project to examine the relationship between eye colour and defects in vision, a total of 878 pupils were classified into five categories on eye colour and independently into three categories according to whether they had normal, short or long sight. The results can be analysed by means of a 3×5 table.

	Normal	Diff.	Short	Diff.	Long	Diff.	(Total)
Blue							
Observed	268	+12.56	75	−9·74	13	−2.81	(356)
Expected	255·44		84·74		15.81		
Green							
Observed	75	−5·36	32	+5·34	5	+0.03	(112)
Expected	80.36		26.66		4·97		
Grey							
Observed	92	−1.28	30	−0.95	8	+2.23	(130)
Expected	93.28		30.95		5·77		
Hazel							
Observed	74	−6.80	33	+5.86	6	+0.94	(114)
Expected	81.80		27.14		5.06		
Brown							
Observed	120	+0.89	39	−1.51	7	−0.37	(166)
Expected	119.11		39·51		7·37		
(Total)	(630)	(0.01)	(209)	(0.00)	(39)	(0.02)	(878)

$$\chi^2 = \frac{12.56^2}{255{\cdot}44} + \frac{9{\cdot}74^2}{84{\cdot}74} + \frac{2.81^2}{15.81} + \frac{5{\cdot}36^2}{80.36} + \text{etc.}$$
$$= 0.6176 + 1.1195 + 0.4994 + 0.3575 + \text{etc.}$$
$$= \underline{6.61} \text{ with } (3-1)(5-1) = \text{eight degrees of freedom}$$

This value corresponds with the tabulated value for a probability of about 0.5. The conclusion, then, is that no significant association between the two classifications has been demonstrated. (For a c × r table with significant association see §8.2.)

7·4 Test of heterogeneity

In biology, the size of an experiment is often limited by availability of living material, extent of culture facilities, or the time available for carrying out the operations involved. When the results of several similar experiments are amalgamated for analysis, it is necessary to test the consistency of the results between experiments to check that they agree well enough to be regarded as being derived from a single experiment. Even when a large experiment or survey can be carried out at one time, if there are doubts about the uniformity of experimental material or of experimental conditions it is an advantage to have data sub-divided so as to provide for a check on technique. We are concerned here only with data in the form of observed frequencies; the parallel situation with measurement data is dealt with in Chapter 10.

There are two kinds of situation with which we have to deal. In one, we are concerned with agreement between the observed frequencies and hypothetical expected frequencies; in the other we are not, but are concerned *only* in testing for heterogeneity. When we are not, we have only to test the hypothesis of no association between two simultaneous classifications and we can use the methods described in §§7.1 and 7.3. We will consider a further example which yields a 3×3 table.

In the course of some work on the autotropism of fungal spores, the germination patterns of touching spore pairs were classified into three categories, thus:

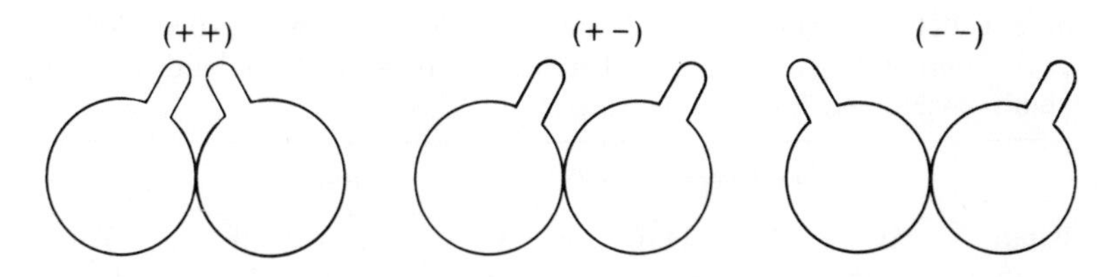

Because of the relationship between the time taken to find, observe and record each spore pair, and the time within which all observations on a particular batch of material had to be completed, the number of spore pairs examined in a single experiment was limited to about 200. In order to provide enough observations for reliable conclusions to be drawn, three such experiments were completed with the following results:

	Germination pattern			
Experiment	(+ +)	(+ −)	(− −)	(Total)
1	36	88	76	(200)
2	32	82	86	(200)
3	40	94	66	(200)
(Total)	(108)	(264)	(228)	(600)

We first calculate the expected values from the products of the marginal totals and as row totals are all equal the expected frequency of (+ +) in all experiments is given by $(200 \times 180)/600 = 36$. Similarly for (+ −) it is 88 and for (− −) it is 76. We then calculate χ^2 as $\Sigma(O-E)^2/E$

$$\chi^2 = \frac{0^2+4^2+4^2}{36} + \frac{0^2+6^2+6^2}{88} + \frac{0^2+10^2+10^2}{76}$$

$$=\frac{32}{36}+\frac{72}{88}+\frac{200}{76}=4.34 \text{ with } (3-1)(3-1)= \text{ four degrees of freedom}$$

This corresponds to a probability of rather greater than 0.3. Deviations from expectation, then, are not significant and data from the three experiments may be totalled for consideration.

7.5 Combined test of heterogeneity and goodness-of-fit

Although the observed frequencies in the above example were used to generate a causal hypothesis, no existing hypothesis was under test when the experiments were carried out. When we have a hypothesis linking the observed frequencies within a single experiment and for the data as a whole, we need to test *both* for agreement with hypothesis and for heterogeneity.

A genetics class examined the *Drosophila* population resulting from crosses between flies heterozygous for a recessive colour mutant. A ratio of 3:1 between wild-type and mutant phenotypes was expected. The population was scored in five batches by individual students and the results assembled for analysis, thus:

	Wild type			Mutant			
Batch	O	E	Diff.	O	E	Diff.	(Total)
1	60	64.5	−4.5	26	21.5	+4.5	(86)
2	75	81.0	−6.0	33	27.0	+6.0	(108)
3	81	74.25	+6.75	18	24.75	−6.75	(99)
4	70	84.0	−14.0	42	28.0	+14.0	(112)
5	54	56.25	−2.25	21	18.75	+2.25	(75)
(Total)	(340)	(360.0)	(−20.0)	(140)	(120.0)	(+20.0)	(480)

For comparisons between the 10 observed frequencies, we have nine degrees of freedom. Four of these could be used in testing agreement of the batch totals with some hypothesis, but these totals are quite arbitrary and such tests would be meaningless. The five remaining degrees of freedom may be used *either* in testing agreement with hypothesis within each of the batches *or* in testing both agreement of population totals with hypothesis and in testing heterogeneity.

We first calculate the χ^2 value for the goodness-of-fit of the observed population totals to those expected on the hypothesis of a 3:1 ratio, thus

$$\chi^2 \text{ value for overall segregation} = \frac{(340-360)^2}{360}+\frac{(140-120)^2}{120}$$

$$=4.444 \text{ with one degree of freedom}$$

This corresponds to a probability of less than 0.05, so significant departure from expectation is indicated.

We next calculate the χ^2 value for the goodness-of-fit of the observed frequencies in each batch to the 3:1 ratio and total these to give the total χ^2 value for the five degrees of freedom, thus:

Batch 1 $\chi^2 = \frac{-4.5^2}{64.5} + \frac{4.5^2}{21.5} = 1.256$

Batch 2 $\chi^2 = \frac{-6.0^2}{81.0} + \frac{6.0^2}{27.0} = 1.778$

Batch 3 $\chi^2 = \frac{6.75}{74.25} + \frac{-6.75}{24.75} = 2.455$ all with one degree of freedom

Batch 4 $\chi^2 = \frac{-14.0^2}{84.0} + \frac{14.0^2}{28.0} = 9.333$

Batch 5 $\chi^2 = \frac{-2.25^2}{56.25} + \frac{2.25^2}{18.75} = 0.360$

Total 15.182 with five degrees of freedom

We then find the χ^2 value for heterogeneity by subtraction and display the results in tabular form:

Sources of variation	Degrees of freedom	χ^2	Probability
Deviation from 3:1 ratio	1	4.444	less than 0.05
Heterogeneity	4	10.738	less than 0.05
(Total)	(5)	(15.182)	—

In the presence of significant heterogeneity, we must question the meaning of the overall deviation from the 3:1 ratio. We re-examine the whole table and note:

(a) The deviation for batch 4 is outstandingly large, with an individual χ^2 value of 9.333 which corresponds to a probability of less than 0.01.

(b) No other batch shows significant deviation, all other χ^2 values being less than 3.841 which corresponds to a probability of 0.05.

Investigation reveals that student No. 4 has defective colour vision and that the sorting of batch 4 is unreliable. As the material of batch 4 has been disposed of it cannot be re-sorted, so the data for this batch are deleted and the remainder re-analysed.

(N.B. The discarding of aberrant data in this way is only justified if evidence other than that provided by the analysis shows it to be erroneous.) (See Problem 7–4.)

7.6 Test of linkage

In the last two sections (7.4 and 7.5) we have met two kinds of situation which may be represented by c × r tables. In one we were interested in testing the agreement of one set of marginal totals with a hypothesis. There is a third kind of situation in which we are interested in the frequency distribution in *both* row and column totals, and in testing for interrelationship (i.e. lack of independence). This arises when we are examining the pattern of segregation of two gene pairs.

Consider the following example. Doubly heterozygous Maize (*Zea mays*) plants were backcrossed with the double recessive. Four kinds of offspring resulted, with the following frequencies:

Red pericarp/not dwarf	PD	204
Red pericarp/dwarf	Pd	153
Colourless pericarp/not dwarf	pD	154
Colourless pericarp/dwarf	pd	165
Total		(676)

On the hypothesis of independent segregation, the expected frequencies would be equal and 169. We first test for goodness-of-fit to this expectation. The total χ^2 value could be calculated as

$$\chi^2 = \frac{(204-169)^2}{169} + \frac{(153-169)^2}{169} + \frac{(154-169)^2}{169} + \frac{(165-169)^2}{169}$$

but as the expected values are the same throughout, we use the simple form

$$\chi^2 = \Sigma(O^2/E) - n$$

$$\chi^2 = \frac{204^2 + 153^2 + 154^2 + 165^2}{169} - 676$$

$$= \underline{10.189} \text{ with three degrees of freedom}$$

This corresponds to a probability of less than 0.02. We thus have clear indication of significant departure from expectation, but we do not know whether this is due to departure from the 1:1 ratio in the segregation P–p, or segregation D–d, or because the two segregations are not independent (i.e. because of linkage), or some combination of these. In order to resolve this, we partition the χ^2 and make *three* comparisons each with one degree of freedom. We may display the data in the form of a 2×2 table, thus:

	P	p	(Total)
D	204 (a)	154 (b)	($a+b$)
d	153 (c)	165 (d)	($c+d$)
(Total)	($a+c$)	($b+d$)	($a+b+c+d=n$)

For the segregation P − p, we test the goodness-of-fit of $(a+c)$ and $(b+d)$ to the ratio 1:1 by calculating:

$$\chi^2 = \frac{[(a+c)-(b+d)]^2}{n} = 2.136$$

For the segregation D − d, we calculate:

$$\chi^2 = \frac{[(a+b)-(c+d)]^2}{n} = 2.366$$

For the linkage, we calculate:

$$\chi^2 = \frac{[(a+d)-(b+c)]^2}{n} = 5.686$$

We may again express the result in tabular form:

Sources of variation	Degress of freedom	χ^2	Probability
Segregation P − p	1	2.136	0.2 to 0.1
Segregation D − d	1	2.366	0.2 to 0.1
Linkage	1	5.686	0.02 to 0.01
(Total)	(3)	(10.188)	—

This leads us to the conclusion that deviation from 1:1 expectation is not significant but that there is significant linkage between P and D and p and d.

When $F_1 \times F_1$ crosses are made, the Mendelian expectation is 9:3:3:1. The same kind of analysis may be carried out, but the expressions used for calculating the individual χ^2 values are as follows (using the same nomenclature as in the last example):

For segregation P − p:

$$\chi^2 = \frac{1}{3n}[(a+c)-3(b+d)]^2$$

For segregation D − d:

$$\chi^2 = \frac{1}{3n}[(a+b)-3(c+d)]^2$$

For linkage:

$$\chi^2 = \frac{1}{9n}[(a+9d)-3(b+c)]^2$$

In our exploration of the use of χ^2 for examining frequency data subjected to more than one classification, we have met an important new idea. We have seen how total variation corresponding to several degrees of freedom can be analysed to permit us to make selected comparisons corresponding to individual degrees of freedom. We do this by *partitioning*

χ^2. In the last part of this book we shall see how to analyse variation in measurement data according to its several causes by partitioning sums-of-squares in the *analysis of variance*.

Problems

7–1 During a study of the small-scale heterogeneity of grazed pastures, casual observation of a number of sites suggested that the presence of White Clover (*Trifolium repens*) was associated with the casting activity of earthworms. In order to test this hypothesis, one site was chosen and a grid of 10×40 contiguous area samples, each one foot square, was marked out. The presence or absence of White Clover (stem and/or leaf) and of fresh worm casts was recorded for each sample. The results were as follows:

	Number of samples
Clover and casts	90
Clover alone	170
Casts alone	25
Neither	115
(Total)	(400)

Using a 2×2 table, test the hypothesis of association.

7–2 Repeat Problem 4–3, using a 2×2 table.

7–3 A large biology class had to be divided for practical work into two parts and taught by different members of staff in two different laboratories. At the time of the examination doubts were raised on the equality of opportunity which had been offered to the two groups of students. Their examination results can be summarized as follows:

	<30	30–39	40–49	50–59	60–69	$\geqslant 70$
Group A	3	8	32	18	12	5
Group B	6	12	22	12	8	3

Examine these results for heterogeneity.

7–4 Complete the analysis for the data of §7.5 after deleting the data for batch 4.

7–5 A student investigated the effect of modifying the vaginal pH of mouse (*Mus musculus*) on the sex ratio of offspring and obtained the following results:

	Male	Female
Untreated, control	34	31
pH 7.0	22	16
pH 4.0	9	21
pH 9.2	21	7

Analyse these data using the method of §7.5.

7–6 An F_2 population of Morning Glory (*Pharbitis*) segregating for two genes Gg and Hh was found to consist of the following: GH 123, Gh 30, gH 27 and gh 21, giving a total of 201.

Examine the segregation of each gene pair and test for linkage between them, using the method of §7.6. What are your conclusions?

Interrelationships of Quantitative Variables 8

8.1 Kinds of information and relationships

There are two main kinds of information that we may have about a series of individuals or samples. One kind consists of measurements, counts, scores, etc., which take the form of a numerical assessment. We call this *quantitative* information. The other kind consists of a classification of the individuals or samples into a series of qualitatively different categories. We call this *qualitative* information. We can examine interrelationships for any of the three possible combinations, thus qualitative/qualitative, qualitative/quantitative and quantitative/quantitative. As we shall see, in practice the situation is more complex because the categories used in a qualitative classification may form a semi-quantitative series. Again, even when a full quantitative (numerical) assessment is not made, it may still be possible to arrange the individuals or samples in order, on the basis of a particular criterion, i.e. to rank them.

8.2 Qualitative/qualitative data

Interrelationship based on qualitative information is usually called *association* and examined by means of a c × r table as described in Chapter 7. We calculate χ^2 with (c − 1)(r − 1) degrees of freedom and this permits us to test the association for significance. The χ^2 value itself is not a good measure of the closeness of association as it depends on n, the total number of observations. If we need such a measure for comparing levels of association for two tables with the same number of degrees of freedom we use χ^2/n, the 'mean-square contingency'. The value of χ^2 does not indicate the nature of the association, e.g. for a 2 × 2 table, whether it is positive or negative. This must be discovered by examination of the table, e.g. by the sign and magnitude of the (O − E) term for each cell.

It is possible, and sometimes necessary, to extend this type of treatment to situations in which the information is semi-quantitative in that the individuals or samples are classified into a series of size classes, e.g. no symptom, slight symptoms, severe symptoms and dead, but the method is often unsatisfactory. Large numbers of observations are required and the interpretation of significant association (when demonstrated) is frequently difficult. An example of the successful application of this semi-quantitative method follows.

During an autecological study of a particular woodland moss, it was considered desirable to examine the relationship between the occurrence of the moss and the depth of the raw humus layer on the soil surface. The moss, where present, was seen to grow either vigorously excluding other species and forming a pure weft, or less vigorously and to contribute along with several other species to a loose mixed weft. It was decided to utilize this natural discontinuity in the behaviour of the moss and to employ the c × r table method. Three classes were recognized: 0 = moss absent; 1 = moss present in mixed weft; 2 = moss present in dense pure weft. The depth of raw humus was arbitrarily scored as: 0 = <1 cm; 1 = 1–2 cm; and 2 = >2 cm. A series of 300 random samples, each 25 cm × 25 cm, were located and records made of moss occurrence and raw humus depth, using the categories defined above. The results were entered into a 3 × 3 table, thus:

Humus		Moss 0	Moss 1	Moss 2	(Total)
0	Observed	36	15	7	(58)
	Expected	27.8	17.8	12.4	
	Difference	+8.2	−2.8	−5.4	
1	Observed	65	37	44	(146)
	Expected	70.1	44.8	31.1	
	Difference	−5.1	−7.8	+12.9	
2	Observed	43	40	13	(96)
	Expected	46.1	29.4	20.5	
	Difference	−3.1	+10.6	−7.5	
(Total)		(144)	(92)	(64)	(300)

$$\chi^2 = \frac{8.2^2}{27.8} + \frac{2.8^2}{17.8} + \frac{5.4^2}{12.4} + \frac{5.1^2}{70.1} + \frac{7.8^2}{44.8} + \frac{12.9^2}{31.1} + \frac{3.1^2}{46.1} + \frac{10.6^2}{29.4} + \frac{7.5^2}{20.5}$$

$$= 2.419 + 0.440 + 2.352 + 0.371 + 1.358 + 5.351 + 0.208 + 3.822 + 2.744$$

$$= \underline{19.065} \text{ with four degrees of freedom}$$

corresponding to a probability of less than 0.001.

Highly significant departure from random expectation was established.

On examining differences between expected and observed values in the individual cells of the table three main tendencies became apparent:

(a) Dense pure wefts tend to occur on a more moderate depth of raw humus.

(b) Loose mixed wefts tend to occur on deep raw humus.

(c) The moss tends to be absent from shallow raw humus.

This result suggested the hypothesis that the moss tends to colonize the soil surface when the remains of other species have accumulated to form a certain depth of raw humus. It then grows vigorously, forming a dense pure weft, but gradually by its own growth it brings about further increase in the depth of the raw humus and renders its habitat less suitable for itself. Its vigour falls and other species invade to produce a mixed weft. Microscopic examination of raw humus profiles supported this hypothesis.

8.3 Qualitative/quantitative data

When investigating relationships between the different ways in which individuals and samples vary, we are frequently called upon to examine the relationship between a qualitative classification on the one hand and the level of some quantitative variable on the other. In ecological work, the qualitative classification may relate simply to the presence or absence of a particular species in a sample, and the quantitative variable either to the quantitative representation of another species or to the level of an environmental factor. The problem here is similar to that presented by the results of a simple experiment with two treatments and yielding measurement data. The method is to sort the individuals or samples into two qualitatively different categories, calculate the mean value of the quantitative variable for each category and then examine the difference between the two means for significance using the t test (or d test if larger numbers are involved). (See Problem 8–1.) The qualitative classification may involve more than two categories, e.g. it may relate to the location of a sample within one of several ecological zones or vegetation types. In such a case the t test is inadequate and the methods of Chapters 9 and 10 are called for.

8.4 Quantitative/quantitative data: correlation

We have seen already (Chapter 2) that the frequency distribution of a single series of measurements made on individuals in a population frequently approximates closely to the normal distribution, and we use the properties of the normal curve in our prediction of probability. If we take measurements of two different properties of the individuals of a population, the data frequently approach a bivariate–normal distribution in which there are two variables (say x and y) and each is normally distributed. Whereas the normal distribution is represented by a normal *curve*, a bivariate–normal distribution is represented by a bivariate–normal *surface*. In examining the correlation between two quantitative variables, we use the properties of such a surface.

Consider a situation in which we have measurements x and y for a series of n individuals. We can display such bivariate data in the form of a 'point' or 'scatter' diagram, using two axes at right angles to represent the two variables. We can include also lines representing the means of the two variables (Fig. 8–1).

We require a measure of the degree to which x and y vary together. One simple measure would be the sum of the products of the joint deviations

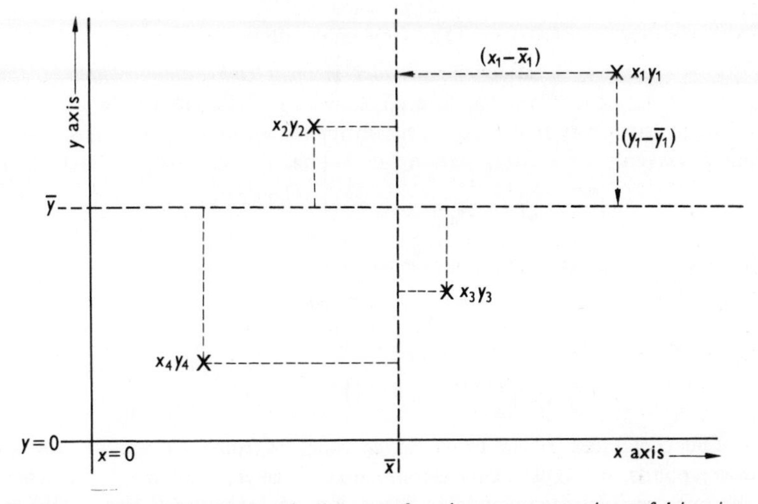

Fig. 8–1 'Point' or 'Scatter' diagram for the representation of bivariate data.

of x and y from their respective means—the 'sums-of-products'—divided by the number of degrees of freedom. Such a measure, a kind of average product, is called the *covariance* of x and y, thus:

$$\text{Covariance of } x \text{ and } y,\ C_{xy} = \frac{1}{n-1}\,\Sigma(x-\bar{x})(y-\bar{y})$$

Each individual contributes to the sum-of-products. In general, some contributions will be positive (points 1 and 4) and some negative (points 2 and 3). Unlike the sum of the squares of the deviations of a variable from its mean—the 'sum-of-squares'—a sum-of-products may be negative, and hence a covariance may be negative. A positive covariance indicates that the two variables tend to vary together (to be positively related); a negative covariance indicates that one tends to increase as the other decreases (to be negatively related).

Covariance is not a suitable measure for use in comparing the closeness of relationship because its magnitude (like that of a simple variance) depends on the units in which the measurements (x and y) are made. In order to free it from this disadvantage, we express the deviations of x and y from their means in terms of standard deviations. So the 'average product' becomes

$$\frac{1}{n-1}\sum\left(\frac{(x-\bar{x})}{s_x}\times\frac{(y-\bar{y})}{s_y}\right) \quad \text{or} \quad \frac{1}{n-1}\,\frac{\Sigma(x-\bar{x})(y-\bar{y})}{s_x s_y}$$

or substituting for s_x and s_y,

$$\frac{\Sigma(x-\bar{x})(y-\bar{y})}{\sqrt{[\Sigma(x-\bar{x})^2\ \Sigma(y-\bar{y})^2]}}$$

This quantity is called the *product–moment correlation coefficient* and our estimate of it (from our limited sample) is designated by $\boldsymbol{r}$.

In the same way that the expression for the sum-of-squares has a form more convenient for computation, so also has the sum-of-products, thus:

$$\Sigma(x-\bar{x})(y-\bar{y}) \equiv \Sigma xy - \frac{(\Sigma x)(\Sigma y)}{n}$$

The expression for $\boldsymbol{r}$ may then be written:

$$\boldsymbol{r} = \frac{\Sigma xy - \dfrac{(\Sigma x)(\Sigma y)}{n}}{\sqrt{\left[\left(\Sigma x^2 - \dfrac{(\Sigma x^2)}{n}\right)\left(\Sigma y^2 - \dfrac{(\Sigma y)^2}{n}\right)\right]}}$$

The coefficient $\boldsymbol{r}$ can have values ranging from $+1$ to -1. $\boldsymbol{r}=+1$ corresponds to a rectilinear relationship (of the form $y=a+bx$) in which the two variables are positively related. $\boldsymbol{r}=-1$ corresponds to a rectilinear relationship in which the two variables are negatively related. Values of $\boldsymbol{r}$ near $+1$ and -1 indicate an approach to a rectilinear relationship, but intermediate values of $\boldsymbol{r}$ are more difficult to interpret. Values of $\boldsymbol{r}$ near zero may arise when there is no real relationship or when there is a real relationship but it is curvilinear. Such values should only be interpreted with the aid of scatter diagrams.

Before using $\boldsymbol{r}$ as a measure of correlation, you should satisfy yourself that you have good reasons to suppose the distribution of the data to be bivariate–normal. Both variables must be continuous (or approximately so) and the frequency distribution of each approach normality, as for a random sample of a normally-distributed population. This means that bivariate data in which one of the variables is controlled, e.g. as time intervals between readings, or as levels of applied treatments, are not eligible. When investigating a certain kind of situation for the first time, it is usually desirable to prepare and examine a scatter diagram before making any calculations.*

* *Practical note.* The computation of $\boldsymbol{r}$ is best carried out in steps with the quantities set down as they are derived, thus:

$\sum x$	$\sum y$	n
$\sum x^2$	$\sum y^2$	$\sum xy$
$\dfrac{(\sum x)^2}{n}$	$\dfrac{(\sum y)^2}{n}$	$\dfrac{(\sum x)(\sum y)}{n}$
$\sum x^2 - \dfrac{(\sum x)^2}{n}$	$\sum y^2 - \dfrac{(\sum y^2)}{n}$	$\sum xy - \dfrac{(\sum x)(\sum y)}{n}$

Hence

$$\boldsymbol{r} = \frac{\sum xy - \dfrac{(\sum x)(\sum y)}{n}}{\sqrt{\left[\left(\sum x^2 - \dfrac{(\sum x)^2}{n}\right)\left(\sum y^2 - \dfrac{(\sum y)^2}{n}\right)\right]}}$$

As part of a study of variation in Timothy grass (*Phleum pratense*), measurements were made of the length of the uppermost leaf (including sheath) and the inflorescence spike, for a random sample of some 30 flowering shoots. The results were as follows (see also Fig. 8–2):

Shoot No.	1	2	3	4	5	6	7	8	9	10
Leaf (cm)	23.4	22.0	25.0	18.1	18.9	20.5	19.1	27.5	21.6	14.3
Spike (cm)	9.8	9.5	12.2	8.3	9.5	9.2	8.5	12.1	10.4	5.5
Shoot No.	11	12	13	14	15	16	17	18	19	20
Leaf	20.8	16.3	23.1	17.4	17.0	26.8	12.5	18.4	16.7	24.0
Spike	10.6	5.5	10.5	7.4	6.8	11.7	4.1	9.3	6.2	11.0
Shoot No.	21	22	23	24	25	26	27	28	29	30
Leaf	24.2	21.2	15.0	20.0	20.1	19.2	21.0	13.0	19.7	26.0
Spike	10.2	9.6	5.0	8.5	9.7	7.0	7.9	4.7	8.3	12.6

For such data it is reasonable to assume a bivariate–normal distribution, so we may calculate r. Let leaf length $= x$ and spike length $= y$. Derived quantities, displayed as in the footnote, may be written as:

602.8	261.6	30
12 552.20	2428.16	5503.44
12 112.26	2281.15	5256.42
439.94	157.01	247.02

$$\therefore\ r = \frac{247.02}{\sqrt{(439.94 \times 157.01)}} = \underline{+0.940}$$

The correlation coefficient provides a measure of association but does not in itself indicate the significance level of any association. When n is small, there is a real possibility of obtaining a value of r deviating markedly from zero because of accidental 'covariation' in the few pairs of values involved. It is usually desirable, then, to test the significance of any deviation of the calculated value of r from zero.

This may be done by evaluating the ratio (estimated value of r)/(standard error of the estimate) and equating this to t with $(n-2)$ degrees of freedom.* The standard error of r depends on r and n only, and equals $\sqrt{[(1-r^2)/(n-2)]}$. So we evaluate $t = r\sqrt{[(n-2)/(1-r^2)]}$ with $(n-2)$ degrees of freedom.

(N.B. In practice, we rarely carry out this test because tables have been prepared which give values of r corresponding to a range of probability levels for different numbers of degrees of freedom.)

Testing the significance of the departure of the value of r estimated in the Timothy grass example (above), we calculate:

* $(n-2)$ degrees of freedom because *two* parameters, both r and its standard error, have been estimated from the data.

Standard error of the estimate $= \sqrt{[(1 - 0.883)/28]} = 0.0644$
$\therefore\ t = 0.940/0.0644 = 14.58$ with 28 degrees of freedom

This is far greater than the tabulated value of 3.674 for $p = 0.001$.

(N.B. Entry of the value 0.94 directly into the table of r leads us to the same conclusion.)

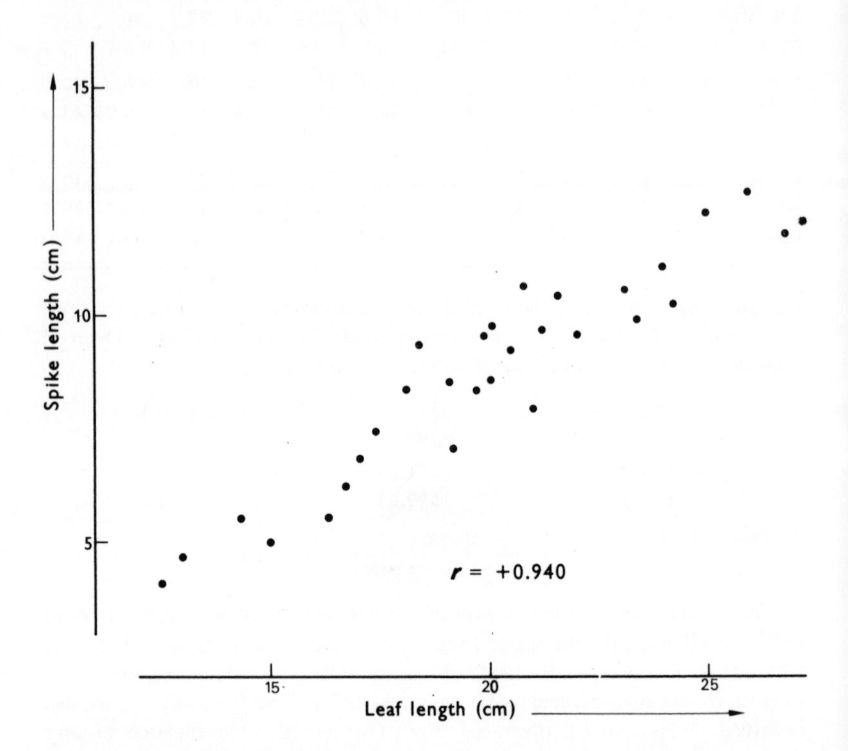

Fig. 8–2 Correlation between leaf length and spike length in Timothy grass (*Phleum pratense*).

8.5 Quantitative/quantitative data: regression

When we examine the relationship between two variables by the method of correlation we make the same assumption about each variable—that it is normally distributed—and we treat the relationship symmetrically. There are many situations in which these assumptions are not justified, e.g. levels of one variable determined by the experimenter. In such circumstances it is more appropriate to examine the relationship by means of *regression analysis*: to define the form of the relationship between the variables, and then to determine how closely the data fit this relationship.

(It is true that in calculating r we are assessing the closeness of a relationship which we assume to be basically rectilinear, but the coefficient tells us nothing about the straight line to which the data tend, except whether its gradient is positive or negative.)

In regression analysis, we recognize the asymmetry of the situation by distinguishing between the independent variable and the dependent variable. The assumptions which we make in the analysis, and therefore the conditions which must be fulfilled before it is applied, are as follows. If x stands for the independent variable and y for the dependent variable, we assume (a) that x can be measured without error, (b) that for every value of x there is a 'true' value of y such as to fit a rectilinear relationship between x and y, (c) that the measurements of y will show random variation and be normally distributed about their 'true' mean, and (d) that the variance of the individual values of y about the 'true' value is the same for all values of x.

The equation of the straight line is of the form $y=a+bx$, where b defines the gradient and a the point at which the regression line crosses the y axis. The values of a and b must be estimated from the data so that the maximum amount of the variation in y is accounted for in terms of variation in x; or in other words, so that the variation in y not accounted for in terms of variation in x is at a minimum. In the same way that we assess variation about a mean in terms of the sum of the squares of the deviations from the mean, so we may assess the departure from linear relationship in terms of the sum of the squares of the deviations from values predicted by the regression equation, i.e. from the regression line (Fig. 8–3).

The regression coefficient b is estimated by:

$$b=\frac{\Sigma(x-\bar{x})(y-\bar{y})}{\Sigma(x-\bar{x})^2}=\frac{\Sigma xy-(\Sigma x)(\Sigma y)/n}{\Sigma x^2-(\Sigma x)^2/n}$$

or alternatively:

$$b=\frac{\text{covariance } xy}{\text{variance } x}$$

The constant a is then estimated by substituting the estimated value of b in the equation $a=\bar{y}-b\bar{x}$. In order to plot the regression line, two convenient values of x are selected (or three to provide a check) and the corresponding values of y are calculated from the regression equation. The regression line should not be drawn beyond the scatter of points which it is calculated to fit.

The application of this procedure will yield a regression line for any set of bivariate data, but where the tendency to linearity is slight the value of b will be very low and regression will account for very little of the variation of the dependent variable. In order to test the significance of

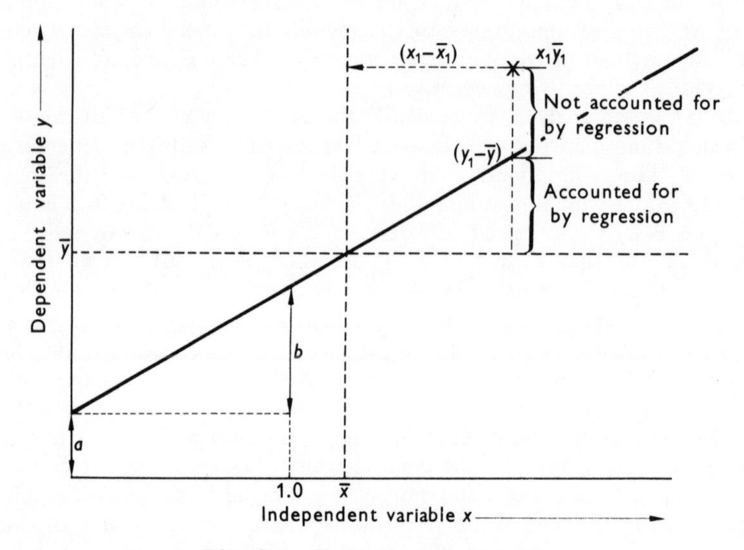

Fig. 8–3 Regression of y upon x.

the regression, it is necessary to analyse the variation of y into two components, as follows:

The total variation of y (as sum-of-squares) is:

$$\Sigma y^2 - \frac{(\Sigma y)^2}{n}$$

The variation accounted for by the regression of y upon x is:

$$\frac{\left(\Sigma xy - \frac{(\Sigma x)(\Sigma y)}{n}\right)^2}{\Sigma x^2 - \frac{(\Sigma x)^2}{n}}$$

So the variation *not* accounted for by the regression, i.e. the residual variation, is:

$$\Sigma y^2 - \frac{(\Sigma y)^2}{n} - \frac{\left(\Sigma xy - \frac{(\Sigma x)(\Sigma y)}{n}\right)^2}{\Sigma x^2 - \frac{(\Sigma x)^2}{n}}$$

This residual sum-of-squares may now be converted to a residual variance by dividing by the appropriate number of degrees of freedom. The variance of y has $(n-1)$ degrees of freedom and we have used one in fitting the regression line (estimating the parameter b) so we are left with $(n-2)$.

$$s_R{}^2 = \frac{1}{n-2}\left[\Sigma y^2 - \frac{(\Sigma y)^2}{n} - \frac{\left(\Sigma xy - \dfrac{(\Sigma x)(\Sigma y)}{n}\right)^2}{\Sigma x^2 - \dfrac{(\Sigma x)^2}{n}}\right]$$

One method of testing the significance of the regression is to test whether the estimated value of b deviates significantly from zero (cf. significance test for $\boldsymbol{r}$). We calculate $\boldsymbol{t} = b(-0)/(\text{standard error of } b)$ with $(n-2)$ degrees of freedom, where the standard error of b is given by $s_R/\sqrt{[\Sigma x^2 - (\Sigma x)^2/n]}$.

Thus:
$\boldsymbol{t} = b\sqrt{[\Sigma x^2 - (\Sigma x)^2/n]}/s_R$ or more conveniently $b\sqrt{[(\Sigma x^2 - (\Sigma x)^2/n)/s_R{}^2]}$.

(N.B. See also §9.5.)

Regression equations are frequently used in the prediction of values for the independent variable (y) for particular values of the independent variable (x). Clearly under such circumstances the accuracy of the prediction becomes important. If the regression equation is based on a relatively large number of samples (i.e. if n is greater than 30), we may ignore any error in fitting the regression line and consider only the variation in y not accounted for by the regression. In such a case, the standard error of the estimated value of y is simply s_R and the 95% confidence limits $y_p \pm 1.96 s_R$ where y_p is the predicted value of y. Such confidence limits may be plotted on to the scatter diagram as two lines parallel to the regression line. (N.B. The standard error of the estimate is the same as the standard deviation because the reliability of a single value is being assessed.)

If the regression equation is based on a small number of samples (e.g. if n is less than 30), there are two sources of uncertainty for which we must allow. One is the estimation of s_R which we meet by replacing the normal deviate (e.g. 1.96 for $p = 0.05$) by the appropriate value of $\boldsymbol{t}$ (with $n-2$ degrees of freedom). The other is in the fitting of the regression line itself. Although we assume that the variance of y about the regression line is the same for all values of x, the accuracy of prediction is not because of the uncertainty in the estimating of b. We allow for this by calculating the confidence limits as

$$y_p \pm \boldsymbol{t} s_R \sqrt{\left(1 + \frac{1}{n} + \frac{(x_p - \bar{x})^2}{\Sigma x^2 - (\Sigma x)^2/n}\right)}$$

where x_p is the value of x for which the predicted value of y is required. Limits calculated in this way become wider as values of x_p deviate from $\bar{x}$.

As part of an investigation on the effect of temperature variation on mice, the rate of water loss by a group of mice was determined for a series of temperatures by absorbing the water evolved by the group in a particular time. The following results were obtained (see also Fig. 8–4):

Temperature (°C)	15	20	25	30	35
Water evolved (mg)	2794	2924	3175	3340	3576

The temperature was determined experimentally and so becomes the independent variable x; the amount of water evolved becomes the dependent variable y. Derived quantities, as displayed in §8.4, may be written as:

125	15 809	5
3375	50 380 213	405 125
3125	49 984 896.2	395 225
250	395 316.8	9900

$$\therefore b = \frac{9900}{250} = 39.60$$

Also $\bar{x} = \frac{125}{5} = 25$ and $\bar{y} = \frac{15\,809}{5} = 3161.8$

$$\therefore a = 3161.8 - (39.60 \times 25) = \underline{2171.8}$$

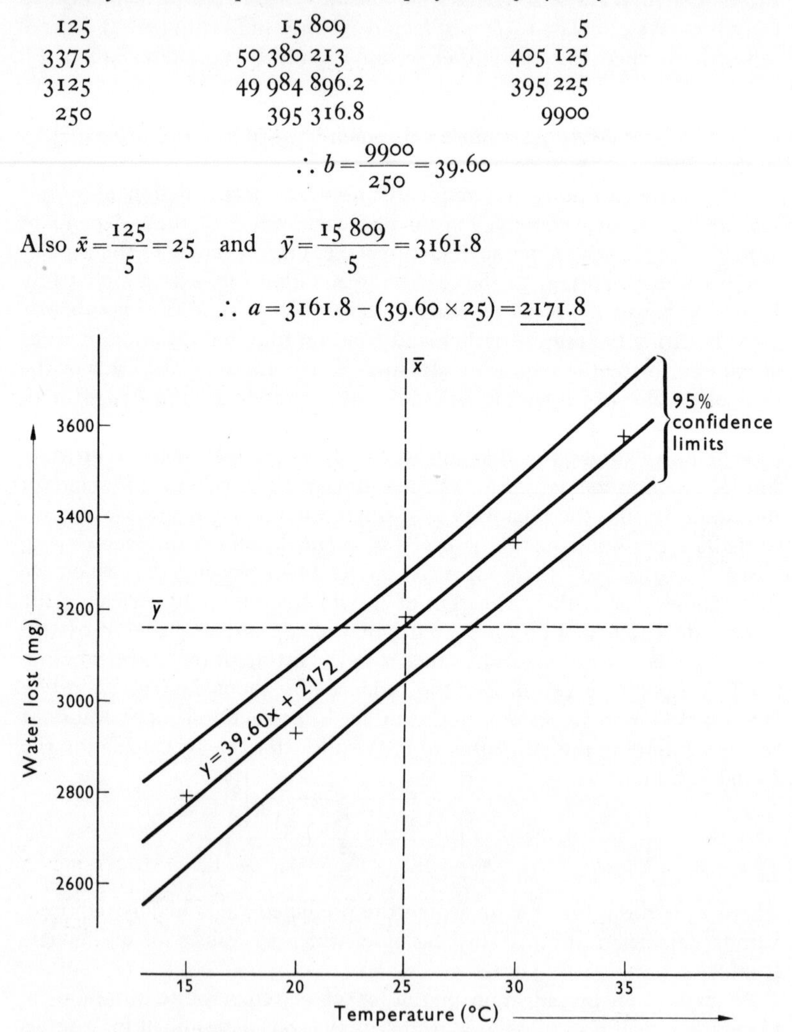

Fig. 8-4 Regression of water loss by mice on temperature, with 95% confidence limits.

Hence the regression equation may be written as $y=39.60x+2172$. To plot the line corresponding to this equation we let $x=15$, then $y=594+2172=2766$, and let $x=35$, then $y=1386+2172=3558$. Plot the points (15/2766) and (35/3558) and join with a straight line.

Calculate the residual variance (s_R^2)

$$s_R^2=\frac{1}{3}\left[395\,316.8-\frac{9900^2}{250}\right]=1092.3$$

Check the significance of the regression by calculating t

$$t=39.6\sqrt{\frac{250}{1092.3}}=\underline{18.945} \text{ with three degrees of freedom}$$

which corresponds to a probability of less than 0.001.

Attach 95% confidence limits (small sample)

$$t_{0.05} \text{ for three degrees of freedom}=3.182$$
$$s_R=\sqrt{1092.3}=33.05$$

Substituting for $x_p=15$ or 35,

$$\text{confidence limits}=y_p\pm 3.182\times 33.05\sqrt{\left(1+\frac{1}{5}+\frac{100}{250}\right)}=y_p\pm\underline{133.0}$$

and similarly for other selected values of x.

8.6 Relationship between coefficients of correlation and regression

The asymmetry of the situations to which regression analysis is normally applied has been stressed. It is possible, however, to calculate *both* the regression of y upon x, and of x upon y. Now the regression coefficient of y upon x is $b_{yx}=\Sigma(x-\bar{x})(y-\bar{y})/\Sigma(x-\bar{x})^2$ and the regression coefficient of x upon y is $b_{xy}=\Sigma(\bar{x}-x)(y-\bar{y})/\Sigma(y-\bar{y})^2$.

So,

$$b_{yx}\times b_{xy}=\left[\frac{\Sigma(x-\bar{x})(y-\bar{y})}{\sqrt{[\Sigma(x-\bar{x})^2\Sigma(y-\bar{y})^2]}}\right]^2=r^2$$

$\therefore\ r=\sqrt{(b_{yx}\times b_{xy})}$ (i.e. the geometric mean of the two regression coefficients). If $r=0$ then the two regression lines are at right angles; if $r=1$ the two regression lines coincide and pass through all the sample points.

8.7 Rank correlation

It sometimes happens that the kinds of variation which we are studying are not susceptible to accurate measurement, although the individuals can be ranked with respect to one another on the basis of two different criteria. In 1904, Spearman, a psychologist, devised a simple form of *rank correlation* applicable to such situations. The Spearman correlation coefficient r_S is evaluated as $r_S = 1 - (6\Sigma d^2)/n(n^2 - 1)$, where d = the difference in ranking order as determined by the two different criteria (it may be positive, negative or zero) and n = the number of individuals. The coefficient r_S like r varies from $+1$ to -1 but the meaning of the two coefficients is not the same, e.g. for $r = +1$ the points must lie on a straight line but for $r_S = +1$ the *ranking order* only for the two kinds of variation must be the same. For values of n greater than 8, the significance of r_S values may be tested exactly as for r, i.e. by means of the t test or by direct reference to tables of r, but for values of n of 8 or less we use the following specially computed table:

Sample size	Probability	
	0.05	0.01
4 or less	none	none
5	1.0	none
6	0.886	1.0
7	0.750	0.893
8	0.714	0.857
9 or more	use table of r	

The method of rank correlation may still be applied when actual measurements are available, e.g. (a) when we have measurements for one kind of variation and ranks for the other and (b) when we have measurements for two variables but the data are not bivariate–normal. Measurements are converted to ranks, and if two or more individuals tie an average rank is accorded to each, e.g. two individuals at equal fifth are both ranked $5\frac{1}{2}$, three individuals at equal eighth are all ranked 9 with the next rank 11.

(N.B. *Non-parametric methods.* As the rank correlation method does not depend for its validity on any assumption concerning the distribution pattern of the variables investigated, it is by definition a non-parametric method. Other methods have been devised which are non-parametric equivalents of the t test and paired-sample test (Chapter 3). Accounts of these and of other non-parametric methods will be found in several of the texts listed on page 116, especially in CAMPBELL, R. C. (1967).

Problems

8–1 It was wished to examine the effectiveness of *Trifolium repens* (White Clover) in accelerating the rate of accumulation of nitrogen compounds in a soil developing from recently exposed subsoil material. An area of suitable made-up ground was selected on which plants of *T. repens* had become established some 3–4 years before and had since developed into distinct colonies about 2–3 ft across. Several transects were marked out and surface soil samples were collected at 1 m intervals along them. The samples were classified into those taken from within clover colonies (T+), and those taken from between the colonies (T−). The soil samples were analysed for total combined nitrogen using the Kjeldahl procedure and the results expressed as mg nitrogen/g dry soil.

	Number	Mean nitrogen content	Variance
T+	15	1.08	0.0190
T−	25	0.96	0.0158

Compare the two mean values for nitrogen content (T+ and T−) using the ***t*** test.

8–2 In a study of the environmental factors controlling the performance of *Ammophila arenaria* (Marram Grass) on a coastal sand dune system, data were collected for total fresh weight of the grass and humus content of the soil for a series of 24 area samples. A point diagram of plant weight against soil humus (expressed in terms of optical density of an alkaline soil extract) was drawn. This indicated a curvilinear relationship and suggested the desirability of transforming the soil humus figures to their logarithms. This was done and a second point diagram plotted. The relationship now appeared to be basically rectilinear. The results were as follows, where $x = \log_{10}$ (optical density of alkaline soil extract) and y = weight in kilograms of *Ammophila* per sample:

$\Sigma x = 38.37$ $\qquad \Sigma y = 7.491$ $\qquad n = 24$

$\Sigma x^2 = 64.344$ $\qquad \Sigma y^2 = 4.328$ $\qquad \Sigma xy = 10.295$

(a) Deduce the equation for the regression of y and x.

(b) Test the significance of the deviation of the regression coefficient (b) from zero, using the ***t*** test.

(c) Calculate the product–moment correlation coefficient (***r***).

(d) Test the significance of the deviation of the correlation coefficient from zero, using the ***t*** test.

8–3 During an investigation of the cause(s) of irregular growth in a plantation of *Picea sitchensis* (Sitka Spruce), 40 trees were selected so as to cover the full height range within a certain plot. Each of the 40 trees was felled, its total height measured and the nutrient content of the needles from the leading shoot determined by chemical analysis. Examine the regression of tree height in metres (y) on nitrogen content of the foliage in % nitrogen per unit dry weight (x). Test the significance of the regression coefficient (b) by means of the t test. Determine the 95% confidence limits of a single estimate of tree height for a particular nitrogen percentage.

Summary of results:

$\Sigma x = 40.92$ $\Sigma y = 249.36$ $n = 40$
$\Sigma x^2 = 46.885$ $\Sigma y^2 = 2038.25$ $\Sigma xy = 297.93$

8–4 Two different methods were used to place a series of 24 vegetation samples in an order which would most effectively illustrate the similarities and differences between them (i.e. to 'ordinate' them). The results were as follows:

Rank:	1	2	3	4	5	6	7	8	9	10	11	12
Ordination A	9	17	5	15	24	13	2	22	11	20	7	19
Ordination B	15	24	9	17	22	20	2	5	13	11	23	1
Rank:	13	14	15	16	17	18	19	20	21	22	23	24
Ordination A	8	23	1	3	14	12	10	16	6	18	4	21
Ordination B	19	7	3	8	18	4	14	6	10	12	21	16

Compare the two ordinations by means of the Spearman rank correlation coefficient.

Analysis of Variance: Single Classifications 9

9.1 Why do we need 'analysis of variance'?

We have considered already (Chapter 3) how the difference between two sample means may be tested for significance. In practice, we are often faced with the problem of examining the differences between the means of *more than two samples*. On meeting this problem for the first time we usually react by testing the significance of each difference separately by means of the ***t*** test. This is neither efficient nor proper.

That it is not proper can be understood by recalling the nature of a test of significance. When we apply such a test at (say) the 5% level of significance, there is a 1:20 probability of our concluding that two population means are different when in fact they are equal (i.e. of making a Type 1 error). This is so for each comparison we make, and it has been estimated that for an experiment with seven treatments, and hence 21 comparisons, there is a probability of 0.67 of making a Type 1 error. This is clearly not good enough. It would be possible by using the 1% level of significance to reduce this probability but to do this would greatly increase the risk of rejecting as non-significant differences which are real (i.e. of making a Type 2 error). To deal with comparisons between more than two sample means a different technique is needed, one which examines the variation within the whole group of sample means—the technique of *analysis of variance*. That the use of multiple ***t*** tests is not efficient will become clear when the economy and elegance of the analysis of variance technique has been appreciated. The analysis is carried out in two steps:

Step 1. We first test the null hypothesis that the samples could have been drawn from a single population, or from several populations with equal means. If we reject this hypothesis, we conclude that the samples represent several populations with different means.

Step 2. We can then proceed to make certain chosen comparisons, e.g. between pairs or groups of sample means.

9.2 Preliminary analysis

Despite its name, analysis of variance does not involve the analysis of variance itself but the *partitioning of the total sum-of-squares*. It consists of a comparison between two estimates of the overall variance (i.e. of the complete set of measurements included in the analysis), one estimate being based on the *variance of the sample means about the grand mean* (the 'between treatment' or 'treatment' variance)* and the other based on the *variance of the individual measurements about their treatment means* (the 'within treatment' or 'error' variance). If the null hypothesis were true we would expect the ratio of these two estimates, (between treatment variance)/(within treatment variance), to approximate to unity. If, on the other hand, the sample means estimate different population means then we would expect the ratio to exceed unity. In practice, we calculate this ratio as the statistic F, and determine the level of probability of obtaining such a ratio (or one larger) if the null hypothesis were true. The two variance estimates are not made independently but as part of a single calculation. It will be recalled that in calculating a simple variance the sum of the squares of the deviations of the individual values from their mean (the sum-of-squares) is divided by the appropriate number of degrees of freedom. In an analysis of variance, we partition the total sum-of-squares according to the cause of variation (e.g. differences between treatments, and random variation between replicates for individual treatments) and divide the several components by their numbers of degrees of freedom to obtain the corresponding variances. Conventionally, these variances are called 'mean-squares'. We then calculate the required variance ratio F as the ratio of two mean-squares.

We have met already, in our use of the statistic $\boldsymbol{t}$, a ratio of the form (mean difference)/(standard error of mean difference) in which we have to take account of the number of degrees of freedom of the denominator. The statistic F is a ratio of two variances, so the numerator and denominator each have their own number of degrees of freedom. We have to take account of both of these when entering a calculated value in the table of F.

In the same way that we use the identity $\Sigma(x-\bar{x})^2 \equiv \Sigma x^2 - \Sigma(\Sigma x)^2/n$ when estimating individual variances, so we use it in the analysis of variance in the calculation of sums-of-squares. It will be seen from the following example that this permits the use of the same 'correction term' throughout.

Method: Analysis of variance: one-way classification

Consider the results of an experiment with k treatments and n replicates for each treatment. The results would have the form:

Treatments

1	2	3	...	k
x_1	x_2	x_3	...	x_k
(n replicates per treatment)				
T_1	T_2	T_3	...	T_k (treatment totals)
$\bar{x}_1$	$\bar{x}_2$	$\bar{x}_3$	...	$\bar{x}_k$ (treatment means)

* The analysis-of-variance technique was introduced in connection with the analysis of data from and the design of agricultural field experiments. Some terms used to describe the analysis reflect this origin. One such term is 'treatment'; others, which will be met later, are 'plot' and 'block'.

Grand total . . . GT and grand mean . . . $\overline{X}$

(i) Calculate the *correction term* C as $(\mathrm{GT})^2/kn$.
(This is the square of the sum of all values of x divided by the total number of values. It is thus equivalent to the $(\Sigma x)^2/n$ term in a simple variance calculation.)

(ii) Calculate the *total sum-of-squares* SS as $\Sigma x^2 - \mathrm{C}$.
(This equals $\Sigma(x - \bar{x})^2$, the sum-of-squares term in a simple variance calculation.)

(iii) Calculate the sum-of-squares for (between) treatments SST as $\Sigma \mathrm{T}^2/n - \mathrm{C}$.
(What is needed here is a sum-of-squares term, based on treatment means or treatment totals, which when divided by the appropriate number of degrees of freedom will yield an estimate of overall variance. The sum-of-squares of the treatment means about the grand mean would be given by $\Sigma(\bar{x} - \overline{X})^2$ but since the sum-of-squares of the individual values is n times that of their means (for means of n) the expression needed is $n\Sigma(\bar{x} - \overline{X})^2$. Replacing means by totals, this becomes $\Sigma \mathrm{T}^2/n - \mathrm{C}$.)

(iv) Calculate the *sum-of-squares for error* SSE as SS − SST.
(We would calculate this directly as:

$$\Sigma(x_1 - \bar{x}_1)^2 + \Sigma(x_2 - \bar{x}_2)^2 + \Sigma(x_3 - \bar{x}_3)^2 + \ldots \Sigma(x_k - \bar{x}_k)^2$$

but it may be shown that this equals SS − SST.)

(v) Prepare the table:

Sources of variation	Sums-of-squares	Degrees of freedom	Mean-squares	F
Treatments	SST	$k - 1$	$\frac{\mathrm{SST}}{k-1}$	$\frac{\mathrm{SST} \times k(n-1)}{\mathrm{SSE} \times (k-1)}$
Error	SSE	$k(n - 1)$	$\frac{\mathrm{SSE}}{k(n-1)}$	—
(Total)	(SS)	$(kn - 1)$	—	—

(vi) Enter the calculated value of F in the table for the level of probability required (e.g. 0.05) at $(k - 1)$ numerator degrees of freedom and $k(n - 1)$ denominator degrees of freedom. If the calculated value is equal to or greater than the tabulated value, we reject the null hypothesis and conclude that the samples have been drawn from populations with different means; or, in other words, that there are significant differences between the treatment means.
(N.B. *Unequal numbers of replicates.* Although experiments are usually planned with equal numbers of replicates per treatment, we often wish to apply analysis of variance to data in which the numbers of replicates are

not equal. A minor modification in the calculation of SST will accommodate this. For numbers of replicates $n_1, n_2, n_3, \ldots n_k$ we calculate SST as:

$$\mathrm{SST} = \frac{T_1^2}{n_1} + \frac{T_2^2}{n_2} + \frac{T_3^2}{n_3} + \ldots + \frac{T_k^2}{n_k} - C = \sum \frac{T^2}{n} - C$$

9.3 Comparisons between sample means

The analysis may be continued in either of two ways:

(a) *Partition of treatment sum-of-squares.* In the preliminary analysis, the treatment sum-of-squares is extracted with a number of degrees of freedom one less than the number of treatments (i.e. $k - 1$). For an analysis with two treatments, there is only one degree of freedom for between treatment comparisons and the F test becomes the equivalent to a t test. When the number of treatments exceeds two, the treatment sum-of-squares itself may be partitioned to permit the testing of a number of comparisons. If the logic of the situation requires it, a whole series of independent comparisons, each with one degree of freedom, can be made. Continuation of the analysis by partition of treatment sum-of-squares may be justified even when the preliminary analysis of variance has failed to show significant overall variation between treatment means.

No attempt will be made here to deal with this kind of development, but an example of partition of treatment sum-of-squares is given as an additional solution to Problem 9–1.

(b) *Testing the significance of individual differences between treatment means.** Several methods are available for examining individual differences between treatment means, but they should not be applied indiscriminately. It is important to understand the problems involved in making multiple tests of significance so that experiments may be properly designed and appropriate methods of analysis selected.

If we apply a test of significance at the 0.05 probability level, there is a 5% risk *for each test* of making a Type 1 error. If an analysis involves us making a whole series of such tests, or if we select for testing pairs of means which show the greatest differences, the risk of making a Type 1 error is much greater. If we now take steps to reduce this risk, say to 5% *for the whole analysis*, the significance level for each individual test is lowered (e.g. to 0.02, 0.01 or lower) and the risk of making Type 2 errors greatly increased.

We can often reduce the severity of this problem by defining precisely the objects of an experiment or survey, deciding on the exact questions to which we require answers and nominating *in advance* the limited number of comparisons that we intend to make. We may then feel justified in

* N.B. The following tests are normally only applied when the preliminary analysis of variance has indicated that there are significant differences between the treatment means.

making several *a priori* comparisons at the usual 0.05 level of significance per test and thus retain the full sensitivity of the analysis. To do this we can use either the modified **t** test or the method of least significant difference (L.S.D.).

Least significant difference—L.S.D.

When applying a **t** test, we calculate **t** as the ratio (mean difference)/(standard error of the mean difference). The true variances of the two populations sampled are assumed to be equal and the standard error used in the test is estimated from the combined data. In the mean-square for error term of the analysis of variance, we have an estimate of the error variance based on the individuals in all samples (i.e. for all treatments). The standard error of the difference between two means of n then becomes $\sqrt{(2s^2/n)}$ where s^2 is the error mean-square.* This estimate may be used when making a nominated comparison by evaluating $t=(\bar{x}_1-\bar{x}_2)/\sqrt{(2s^2/n)}$ with $k(n-1)$ degrees of freedom.

For a two-treatment analysis, this will give a result identical to that for an ordinary **t** test, but where there are more than two treatments it will tend to be more sensitive because of the larger number of degrees of freedom: $k(n-1)>(n_1+n_2-2)$.

Instead of calculating **t** for a particular pair of means, we can calculate the least difference between a pair of means which would be significant as L.S.D. $=t_{0.5}\sqrt{(2s^2/n)}$ where **t** has $k(n-1)$ degrees of freedom. Pairs of means which differ by more than the L.S.D. may be regarded as significantly different at the probability level of **t** employed. Having calculated the L.S.D., it is very tempting to apply it to differences between all pairs of means, or pairs selected *a posteriori*. Such a temptation should be resisted for the reasons already stated.

Circumstances often arise in which we need to make a larger number of comparisons between treatment means and/or to nominate comparisons *a posteriori*. In such circumstances, we need to modify the testing procedure to reduce the risk of making Type 1 errors. A simple form of test is the fixed-range test in which we compute a single *least significant range* or L.S.R. using a tabulated multiplier Q. It gives greater protection than the L.S.D. method against Type 1 errors at the expense of an overall loss of sensitivity (i.e. an increased risk of Type 2 errors).

Fixed-range test—using L.S.R.

We can re-state the expression for the L.S.D. as L.S.D. $=t_{0.05}\sqrt{2}s/\sqrt{n}$, and then see that to obtain the L.S.D. we multiply the standard error of an individual mean by $t_{0.05}\sqrt{2}$. Taking the whole analysis rather than the single comparison as the unit, and assuming that all possible comparisons between pairs of treatment means will be made, a multiplier $Q(\geqslant t\sqrt{2})$ has been devised for use instead of $t_{0.05}\sqrt{2}$, which reduces the probability

* This takes the form $\sqrt{(s^2/n_1+s^2/n_2)}$ when $n_1 \neq n_2$.

of making a Type 1 error *per analysis* to the probability level of Q employed, e.g. 0.05.

We calculate the L.S.R. as $Q\sqrt{(s^2/n)}$ where Q is read from the table of the 'Studentized range' for the number of treatments in the analysis and for the number of degrees of freedom for error (i.e. $k(n-1)$). The L.S.R. is then used in the same way as the L.S.D., but of course may be used more freely.

Several methods have been devised which, while providing reasonable protection against Type 1 errors, allow the retention of greater sensitivity. In these the means are arranged in order of magnitude and a series of *shortest significant ranges*, or S.S.R., is calculated for testing differences between pairs of means according to their relationship in the size order.

Multiple-range test—using S.S.R.

As in the fixed-range test, we first compute the standard error of a treatment mean, $\sqrt{(s^2/n)}$. We then look up values of Q (in the table of 'Studentized range') for the number of degrees of freedom for error, and for $k, k-1, k-2, k-3 \ldots, 2$ treatments. We then compute the S.S.R. for groups of $k, k-1, k-2, k-3 \ldots, 2$ as $Q\sqrt{(s^2/n)}$. Having arranged the treatment means in order of magnitude, we test the total difference between the largest and smallest mean by comparison with the S.S.R. for k means. If this is non-significant, we stop testing. If this difference is significant, we test the differences between the largest and next-to-smallest mean, and the next-to-largest and the smallest mean, by comparison with the S.S.R. for $k-1$ means. We continue in this way for as long as differences prove significant, but if a difference proves to be non-significant we do not test any differences within it.

9.4 Assumptions underlying the analysis: transformation of data

Before dealing with further applications of the analysis-of-variance technique, it is desirable to stop and consider the assumptions on which the technique is based so that it may be correctly applied. The two most important assumptions are:

(a) That the effects are additive, i.e. an individual value (of x, say) is considered to be made up of the grand mean + treatment effect + uncontrolled error.

(b) That the uncontrolled error is normally distributed and has an equal variance for all treatments.

Fortunately these assumptions are justified for a great deal of both experimental and survey work and so conclusions drawn from the analyses of the resulting data are valid. Not uncommonly, however, these assumptions are not fully justified and adjustments are necessary. Adjustments frequently take the form of transformation, which consists of converting the original measurements and counts to some mathematical function of them before analysis. The need for transformation is usually apparent on inspection of the data but tests for inequality of variances are available and will be found described in the textbooks. Full treatment of the subject will not be attempted here but a few examples will be given:

(i) Results in the form of small, whole-numbered counts tend to have a variance proportional to their mean (or equal to it, as for the Poisson distribution). An appropriate transformation to render the variances independent of the means is $\sqrt{x}$, or if the counts are generally very low and zeros occur, $\sqrt{(x+0.5)}$.

(ii) Correlations between treatment means and treatment variances are not confined to small whole-numbered counts and more generally the log x transformation is employed, or log $(x+1)$ if zeros are present. The log x transformation is also appropriate when responses tend to be proportional rather than additive, e.g. for treatments at levels 0, 1, 2, 4 yielding treatment means of 0, 10, 20, 40 rather than 10, 12, 14, 18.

(iii) Data in the forms of probabilities, proportions and percentages tend to be binomially distributed. But when the corresponding values of p range only from 0.0–0.2, a Poisson distribution is approached and a $\sqrt{x}$ or $\sqrt{(x+0.5)}$ transformation is used. Values of p of 0·8–1.0 are subtracted from 1.0 (or percentages subtracted from 100%) and treated similarly. Where the range of p is 0.5 ± 0.2, the distribution is almost normal and transformation is not necessary. Where, however, the range of p is greater, the 'angular' or 'arcsin $\sqrt{p}$' transformation should be used. Tables give angles in degrees such that the sine of the angle equals $\sqrt{p}$.

Although arguments for transforming data are statistical, it is often of value to enquire why the data need transformation. To do so may indicate how the data may have been collected in such a way as to not require transformation, or even throw some light on the phenomena which give rise to them. For example, it has been found that the areas of fungal colonies require transformation before analysis, but the radii of colonies do not. This suggests that the important variable here is the distance of spread from the point of inoculation or infection. Similarly, total weights of plants and animals often prove unsatisfactory variates. The fact that log (weight) proves to be satisfactory suggests that the quantity which is responding to treatment is growth rate.

9.5 Analysis of variance in testing the significance of regression

We have seen (Chapter 8) that in regression analysis the overall variation of the dependent variable (y, say) about its mean is considered as made up of two parts, that accounted for by regression and that not accounted for. When testing the significance of regression, we can often do the job more neatly by using F and the analysis-of-variance technique, rather than by using t in testing the deviation of b, the regression coefficient, from zero. We partition the total sum-of-squares and divide each component by its number of degrees of freedom to obtain mean-square terms. We then calculate F = (mean-square for regression)/(residual mean-square). If the calculated value is equal to or more than the tabulated value, then we conclude that the regression is significant.

The sums-of-squares needed in the analysis are:

Total sum-of-squares, $\Sigma y^2 - \dfrac{(\Sigma y)^2}{n}$ with $n-1$ degrees of freedom

Regression sum-of-squares, $\dfrac{\left[\Sigma xy - \dfrac{(\Sigma x)(\Sigma y)}{n}\right]^2}{\Sigma x^2 - \dfrac{(\Sigma x)^2}{n}}$ with one degree of freedom

Residual sum-of-squares by difference, with $n-2$ degrees of freedom

An application of this form of analysis is required in Problem 9–5.

Problems

9–1 In an investigation of the effect of NaCN (sodium cyanide) on the uptake *in vitro* of a particular amino acid by intestinal preparations from a certain species of fish, it was found that each fish would give only about six preparations. Since it would be necessary to use more than one fish in each experiment, a preliminary test was carried out to examine the variation between preparations from different fish. Three preparations were made for each of four fish.

The results obtained, expressed as μmol g^{-1} dry weight per 20 min period, were as follows:

	Fish			
Replicate	1	2	3	4
(i)	2.53	2.02	1.66	1.36
(ii)	2.04	1.92	1.92	1.15
(iii)	2.34	2.03	1.47	1.16

Analyse these data and report on the relationship between the variance due to differences between fish and that due to differences between replicate preparations from the same fish.

9–2 As a result of a discussion on university entrance requirements and, in particular, the relationship between performance in G.C.E. A-level biology and in first year at university, the performances of 120 students were examined. The students were classified into seven groups according to school experience and performance, and their marks in the first-year university examination in botany analysed. The data are summarized below:

	A-level grade					O-level only	No biology at school
	A	B	C	D	E		
n	7	21	23	23	23	8	15
Σx	338	1031	1097	1034	1085	397	684
Σx^2	17 096	53 297	54 305	50 050	52 783	19 965	32 740

Test the null hypothesis that marks in the first-year botany examination are independent of past experience and performance in biology.

9–3 During a study of the photoperiodic control of reproduction in the red alga *Porphyra*, an experiment was conducted to examine the effect of interrupting a long dark period with 30 min illumination by light of different wavelengths but equal energy levels. Ten replicates were used for each of five wavelength bands and the results were recorded in the form of sporangial counts for a standard volume of algal material. It was known that for data of this kind errors in sampling and counting (standard deviation of replicate counts) tended to be proportional to number counted, so logarithmic transformation was called for. Only red light was expected, on theoretical grounds, to have an effect on sporangial production, so it was decided *before the experiment was set up* that the mean for the red-light treatment would be compared with each of the other treatments.

The results obtained were as follows:

Colour	$\bar{x}$	n	$\overline{\log x}$	$\Sigma(\log x)$	$\Sigma(\log x)^2$	$[\Sigma(\log x)]^2$
Blue	7728	10	3.885	38.846	150.926	150.903
Green	7949	10	3.890	38.905	151.447	151.359
Yellow	7107	10	3.842	38.429	147.745	147.675
Red	4759	10	3.604	36.037	130.566	129.869
Far-red	7489	10	3.866	38.657	149.511	149.436
(Totals)		(50)		(190.874)	(730.195)	(729.242)

Complete the analysis of variance and compare the mean for red light with each of the other treatment means using (a) the method of L.S.D. and (b) the fixed-range test and L.S.R.

9–4 As part of the study of the factors affecting the performance of *Juncus squarrosus* (Heath Rush), 10 sites were selected and 12 random area samples (each 0.5 m × 0.5 m) were located in them. The plants of

(Problem **9–4**. Results of potassium determinations, in mg g^{-1})

Batch No.	Sites										Batch totals
	1	2	3	4	5	6	7	8	9	10	
1	10.0	13.6	11.9	11.3	15.4	12.8	14.0	13.8	14.4	18.6	135.8
2	10.0	10.4	11.1	11.9	16.4	13.6	15.3	14.6	13.9	20.2	137.4
3	13.7	12.5	12.4	11.4	13.7	13.1	15.5	13.7	14.1	19.8	139.9
4	13.8	11.9	10.1	12.4	14.9	13.5	16.5	15.8	15.2	19.7	140.8
5	15.4	12.1	11.4	15.2	15.6	14.6	16.6	16.3	14.4	19.7	151.3
6	12.8	14.0	12.0	13.3	16.5	12.9	16.2	14.4	16.4	17.9	146.2
7	12.3	12.0	14.7	15.6	16.9	13.6	15.0	15.9	17.0	18.7	151.7
8	11.4	10.0	14.6	16.5	16.2	14.4	15.2	14.5	15.0	20.0	147.7
9	12.9	9.3	13.2	13.0	17.4	13.4	16.1	13.2	14.0	20.0	142.5
10	8.8	10.0	14.8	14.0	15.8	15.0	16.2	14.2	14.0	21.8	144.6
11	8.2	9.2	13.5	13.4	18.1	11.9	16.0	16.1	13.8	20.2	140.4
12	9.6	10.6	16.1	12.0	16.4	12.3	16.9	14.3	14.3	19.8	142.3
T	138.9	135.6	155.8	160.0	193.3	161.1	189.5	176.8	176.5	236.4	1723.9
x	11.58	11.30	12.98	13.33	16.11	13.43	15.79	14.73	14.71	19.70	$\bar{X}$14.37
			GT=1723.9			$\sum x^2$=25 640.45					

J. squarrosus on each sample area were dug up and after removal of the dead parts were dried, ground up and analysed for a number of nutrient elements. In order to prevent the mean values for nutrients per site being influenced by any systematic errors in the chemical analyses, the analyses were carried out in 12 batches, each batch including one determination for each site. The results of the potassium determinations (in mg g^{-1} dry weight) are given above.

There was no logical basis for the nomination of individual comparisons but two questions were asked:

(i) Are the differences between the site means significant at the 5% level, and if so . . . ?

(ii) Into what groups do the sites fall on the basis of the potassium content of their *J. squarrosus*?

Examine these data by means of analysis of variance and if between-site differences are shown to be significant use the multiple-range test and S.S.R. for grouping the treatment means.

9–5 Re-examine the significance of the regression of Problem 8–2 by means of analysis of variance.

Analysis of Variance: Double Classifications 10

10.1 Experimental design

We have so far been learning how to deal with numerical data gathered from experiments or surveys without paying much attention to the design of the experiments or surveys themselves. This could be misleading because plans for collecting data and analysing them should be made at the same time, as two parts of a single logical operation of testing a hypothesis. This point is perhaps best understood by examining the relationship between experimental design and analysis of variance.

The simplest kind of experiment is one in which we keep constant all factors other than the one to be varied experimentally. Many physical and chemical experiments do approximate to this ideal, but in biology uncontrolled variation in material, handling and environment occurs and leads to variation in the results which has to be accepted as error. There are three essentials in dealing with such a situation:

(i) Reduce individual and sample errors to a minimum.

Methods of reducing individual errors relate to uniformity of experimental material and constancy of conditions, and will depend on the nature of the situation examined. Proportional errors in counts and errors in sample means may both be reduced by increasing the sizes of the samples.

(ii) Avoid bias (i.e. the association of a particular kind of error with a particular treatment).

We avoid bias by the use of *randomization*, i.e. by assigning treatments to individuals by a method which gives each individual an equal chance of being assigned to each treatment. This is normally done with the aid of sets of random numbers.

(iii) Arrange to be able to distinguish variation due to error from variation due to treatments.

We have seen (Chapter 9) how we can separate the effects of treatments from the error, and how we use the ratio of the two corresponding variances to test the significance of the former.

The experimental design corresponding to this kind of analysis is the *fully-randomized* design. Consider this in the form of an agricultural field experiment. We wish to compare the effects of four treatments A, B, C and D. To provide an estimate of error we need replicates. Considering the size of plot necessary and the resources at our command, we decide to use six replicates per treatment. Our layout might be:

C	D	B	D	C	A
A	C	A	B	D	B
B	B	D	D	A	C
D	A	C	C	B	A

i.e. four treatments and six replicates arranged in 24 plots.

The analysis-of-variance table would have the form:

Sources of variation	Sums-of-squares	Degrees of freedom	Mean-squares	F
Treatments		3		
Error		20		—
(Total)		(23)	—	—

One of the sources of error in such an experiment would be the natural variation in soil condition within the area occupied by the experiment. Attempts to improve the sensitivity of the experiment by increasing the number of replicates often lead to an increase in error variance between plots because of the difficulty of finding a large enough area with a uniform soil condition. This difficulty may be overcome by using the *randomized-block* design. In this we locate our plots in several *blocks*, each block containing an equal number of plots subjected to each treatment, the treatments being randomized in the blocks. The layout might be:

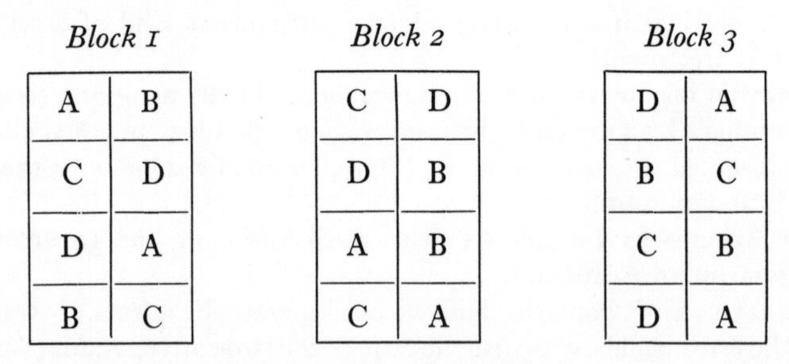

i.e. three blocks each containing eight plots with four treatments and two replicates per block.

The blocks must still be located as far as possible in areas of uniform soil condition but the problem of site selection is reduced because it is easier to find a number of small areas of uniform fertility than one large one.

In the analysis, the variation between blocks is extracted as a separate term so that the error variance, with which the between-treatments variance is compared, is based solely on variations within blocks.

10.2 The method

Consider the results of an experiment with k treatments, b blocks and n replicates per treatment per block. The data would have this form:

	Treatments					
	1	2	3	. . .	k	*Block totals*
Block 1	x_{11}	x_{21}	x_{31}	. . .	x_{k1}	B_1
	(n replicates/treatment)					
Block 2	x_{12}	x_{22}	x_{32}	. . .	x_{k2}	B_2
	(n replicates/treatment)					
Block 3	x_{13}	x_{23}	x_{33}	. . .	x_{k3}	B_3
	(n replicates/treatment)					
	.	.	.		.	
	.	.	.		.	
	.	.	.		.	
	.	.	.		.	
Block b	x_{1b}	x_{2b}	x_{3b}	. . .	x_{kb}	B_b
	(n replicates/treatment)					
Treatment totals	T_1	T_2	T_3		T_k	Grand total GT
Treatment means	$\bar{x}_1$	$\bar{x}_2$	$\bar{x}_3$		$\bar{x}_k$	Grand mean $\bar{X}$

(i) Calculate C $=(\text{GT})^2/bkn$ (there are now bkn values of x)
(ii) Calculate SS $=\Sigma x^2 - \text{C}$ (as before)
(iii) Calculate SST $=\Sigma \text{T}^2/bn - \text{C}$ (there are now bn values of x for each treatment)
(iv) Calculate SSB $=\Sigma \text{B}^2/kn - \text{C}$ (there are now kn values of x for each block)
(v) Calculate SSE $=\text{SS} - (\text{SST} + \text{SSB})$
(vi) Prepare the table:

Sources of variation	Sums-of-squares	Degrees of freedom	Mean-squares	F
Treatments	SST	$k-1$	$\dfrac{\text{SST}}{k-1}$	$\dfrac{\text{SST}}{\text{SSE}} \times \dfrac{(bkn-k-b+1)}{(t-1)}$
Blocks	SSB	$b-1$	$\dfrac{\text{SSB}}{b-1}$	$\dfrac{\text{SSB}}{\text{SSE}} \times \dfrac{(bkn-k-b+1)}{(b-1)}$
Error	SSE	$bkn-k-b+1$	$\dfrac{\text{SSE}}{bkn-k-b+1}$	—
(Total)	(SS)	$(bkn-1)$	—	—

The variance ratio F, for treatments, is calculated as the ratio of the mean-square for treatments over the error mean-square. The variance ratio for blocks may also be calculated, as the ratio of the mean-square for blocks over the error mean-square.

The value of the randomized block design has been considered only in relation to the field experiment. In biology, there are many sources of variation which may be dealt with in the same way as soil fertility. Variation in experimental organisms (or other material) can be largely eliminated from error by forming blocks of groups of organisms selected as being closely similar in age, size, genetic constitution and environmental history (cf. paired-sample test). Again, the effects of variation in environmental conditions to which experimental material is submitted may often be greatly reduced by placing the samples (plots) of one block together, e.g. on the same greenhouse bench or in the same incubator. Once blocks have been defined (for any particular reason), it is desirable to associate with them as many other sources of variation as possible, e.g. whole blocks should be recorded at one time so that any effect of recording time is eliminated from the results for treatments, and the recording of a block should be done by a single observer, so that error is not inflated by differences between observers.

(N.B. The data from a randomized block experiment could be analysed without taking account of the blocks, i.e. using the one-way classification method of Chapter 9. It is instructive to compare the results from the two kinds of analysis. The two-way classification analysis transfers from error to blocks not only the sum-of-squares for blocks but also $(b-1)$ degrees of freedom. Thus, if the variation between blocks is larger than would be expected from the variation between replicates within blocks, a useful reduction in the error mean-square will result. But if there is little difference between blocks, then the fall in number of degrees of freedom for error might result in the actual *increase* in the error mean-square. This is unlikely to be important as long as there are several replicates per treatment per block.)

Interpretation of results

With two-way classification as described above, there are two tests of significance and therefore *four* kinds of possible result:

(i) *Treatments—non-significant: Blocks—non-significant*

This result shows that all values (of x, say) might have been drawn from the same normal population, and no further analysis is called for.

(ii) *Treatments—significant: Blocks—non-significant*

This result shows that the expected variation between blocks has not materialized. Individual differences between treatment means may be tested as for one-way classification, but using the new error mean-square as the error variance estimate.

(iii) *Treatments—significant: Blocks—significant*

This result confirms the existence of between blocks variation and justifies the use of the randomized block design. If the main object of the experiment was to compare treatment means, tests may be carried out as for one-way classification. If, however, quantitative statements are to be made about the effects of particular treatments, e.g. treatment A increases the yield by $Y \pm z$ kg over control, further analysis of variance, as described in the next section, is desirable.

(iv) *Treatments—non-significant: Blocks—significant*

The results are indecisive and further analysis of the data, as described in the next section, is called for.

10.3 Interaction

We have been assuming so far that any effects of treatments and blocks are additive. We have assumed that the value for each plot is the sum of overall mean + treatment effect + block effect + uncontrolled error for that particular plot. This is the same as assuming that treatments have the same effect in all blocks and that differences associated with blocks affect the results of all treatments equally. Any deviation of the means from this additive pattern has been included along with the variation between replicates within plots as error.

It very often happens that effects are not simply additive; responses to treatments differ among the blocks and/or the differences associated with blocks are not the same for each treatment. Such departure from the simple additive pattern is called *interaction*. We need to be able to analyse into two components the term which we have been regarding as error; one relating to the variation between replicates (within treatments, within blocks) and the other relating to the variation in the means of these replicates which cannot be accounted for on the assumption that the effects of treatments and blocks are additive (i.e. interaction). This is, of course, only possible when we have several replicates per treatment per block. With only one such replicate, error is estimated entirely in terms of deviation from the additive pattern.

Consider an experiment with k treatments, b blocks and n replicates per treatment per block. In addition to T representing treatment totals and B representing block totals, we shall need Q (say) representing totals for the groups of n replicates within treatments within blocks (i.e. within the $b \times k$ sub-classes in the table).

(i) Calculate C $= (\mathrm{GT})^2/bkn$

(ii) Calculate SS $= \Sigma x^2 - \mathrm{C}$

(iii) Calculate SSQ $= \Sigma \mathrm{Q}^2/n - \mathrm{C}$ (there are n replicates for each sub-class)

(iv) Calculate SSE $= \mathrm{SS} - \mathrm{SSQ}$ (this estimates the variance within groups of n replicates)

(v) Calculate SST $= \Sigma T^2/bn - C$
(vi) Calculate SSB $= \Sigma B^2/kn - C$
(vii) Calculate SSTB (sum-of-squares for interaction)
$= SSQ - (SST + SSB)$
(viii) Set up the table:

Sources of variation	Sums of-squares	Degrees of freedom	Mean-squares	F
Treatments	SST	$k - 1$	$\frac{SST}{k-1}$	$\frac{SST}{SSF} \times \frac{kb(n-1)}{k-1}$
Blocks	SSB	$b - 1$	$\frac{SSB}{b-1}$	$\frac{SSB}{SSE} \times \frac{kb(n-1)}{b-1}$
Interaction	SSTB	$(b-1)(k-1)$	$\frac{SSTB}{(b-1)(k-1)}$	$\frac{SSTB}{SSE} \times \frac{kb(n-1)}{(b-1)(k-1)}$
Error	SSE	$kb(n-1)$	$\frac{SSE}{kb(n-1)}$	—
(Total)	(SS)	$(kbn - 1)$	—	—

(ix) First calculate F for interaction as mean-square for interaction over error mean-square and enter it in the table at $(b-1)(k-1)$ numerator and $kb(n-1)$ denominator degrees of freedom. If interaction proves to be non-significant, the significance of differences between treatment means and block means may be tested as before but using the new mean-square for error. If the interaction proves to be significant, quantitative statements about the effects of treatments should be made with care. A table of sub-class totals (Q values) or sub-class means (Q/n values) should be drawn up and studied for meaning. If the interaction is small compared with the treatment effects, it may be better to make inclusive statements about the treatment effects using the error mean-square with $(bkn - k - b + 1)$ degrees of freedom (from the previous analysis) to fix confidence limits.* But, if the interaction is large, it will probably be better to regard the experiment as several small experiments, one for each block, and re-analyse accordingly.

10.4 Factorial experiments

We have so far considered analysis of variance only as an aid to making multiple comparisons between treatment means. In the fully-randomized

* If the sum-of-squares for interaction is *not* extracted, both it and the corresponding number of degrees of freedom are included under error.

design, we compare variation due to treatments with that for replicates within treatments. The randomized block design can increase the sensitivity of experiments, and hence their efficiency, by associating in blocks sources of error which can be anticipated but not eliminated. The resulting variation can then be distinguished and separated from that due to treatments on the one hand and the residual error on the other.

If we are concerned with comparisons between treatment means, only the simpler form of analysis (without extraction of mean-square for interaction) may suffice. Where such an analysis shows differences between treatment means to be significant,* the means may be grouped and/or have confidence limits attached without analysing the total sum-of-squares further. If, however, the differences between treatment means is non-significant but that for differences between blocks is significant,† it is often desirable to know whether this result is due to absence of real differences between treatment means of because the error mean-square has been inflated by interaction. Under these circumstances it is desirable to examine the data for interaction. If significant interaction is demonstrated, the hypothesis of additivity (independence of effects) must be rejected and assumptions on which the experiment was based re-examined. It might be necessary for the nature of the interaction to be investigated; if not, at least the design of further experiments would need to be modified.

Interaction is not always unexpected and unwanted. In some fields of enquiry the investigation of interaction is an important aspect of the work. The interactions between the different nutrient elements applied in fertilizer experiments may be as important as the effects of the individual elements. In its simplest form, such work requires no modification in the form of analysis. The several levels of one variable are treated as 'treatments' and the several levels of the other as 'blocks'. The data may be analysed to yield meaningful conclusions relating to both main effects (variation in individual factors) and to interaction. Such experiments, known as *factorial experiments*, have been developed to a high state of sophistication and the complete analysis of data resulting from the more advanced designs would take us well beyond the scope of this book.

10.5 Samples within samples

We have learnt how to analyse data from several kinds of simple experiment or survey in which the population samples consist of similar units each contributing a single measurement, e.g. individual plants each contributing a height or oven-dry weight. Not infrequently the individual, by necessity or design, contributes a number of measurements, each for a separate sub-sample. Such sub-sampling raises important questions of

* Result (ii) on p. 80 and (iii) on p. 81.
† Result (iv) on p. 81.

efficiency, and before it is employed it is desirable to investigate the variation corresponding to each of the two or more sampling levels. An example with three levels of sampling will illustrate this.

A student proposed to investigate the effect of different light intensities on the mean chloroplast number per cell in the leaves of the moss *Funaria hygrometrica*. Decisions were needed on the number of cells to be counted per leaf, the number of leaves to be examined per plant sample and the number of samples to be examined for each light intensity. To facilitate these decisions the chloroplasts in each of ten cells in each of five leaves for each of four plant samples, all gathered within a single light intensity, were counted.

The results obtained were as follows:

	Sample 1					*Sample 2*				
Leaf	1	2	3	4	5	1	2	3	4	5
Σx	134	126	127	135	132	145	130	148	139	140
Σx^2	1814	1622	1625	1867	1792	2141	1712	2198	2039	1992
	$\Sigma x_1 = 654$					$\Sigma x_2 = 702$				

	Sample 3					*Sample 4*				
Leaf	1	2	3	4	5	1	2	3	4	5
Σx	146	147	143	152	149	140	121	134	146	145
Σx^2	2174	2167	2079	2314	2239	1980	1495	1818	2132	2188
	$\Sigma x_3 = 737$					$\Sigma x_4 = 686$				

$$\Sigma x \text{ (sample)} = 2779$$
$$\Sigma x^2 \text{ (sample)} = 39\,388$$

We analyse the total sum-of-squares (for the 200 counts) into components (a) between samples, (b) between leaves within samples and (c) between counts within leaves.

$$C = \frac{2779^2}{200} = 38\,614.21$$

$$\text{SS counts} = 39\,388 - C = 773.79$$

$$\text{SS leaves} = \frac{134^2 + 126^2 + 127^2 + 135^2 + 132^2 + \ldots + 145^2}{10} - C$$

$$= 143.49$$

$$\text{SS samples} = \frac{654^2 + 702^2 + 737^2 + 686^2}{50} - C = 71.49$$

Then, by subtraction:

$$\text{SS between leaves within samples} = 143.49 - 71.49 = 72.00$$
$$\text{SS between counts within leaves} = 773.79 - 143.49 = 630.30$$

Sources of variation	Sums-of-squares	Degrees of freedom	Mean-squares	F
Between samples	71.49	3	23.83	5.30
Between leaves within samples	72.00	16	4.50	1.29
Between counts within leaves	630.30	180	3.50	—
(Total)	(733.79)	(199)	—	—

The number of degrees of freedom for examining variation between the four samples is $4-1=3$. The total number of degrees of freedom for examining variation between the 20 leaves is $20-1=19$, but this number includes the three already being used for between-sample comparisons, so the number available for examining variation *between leaves within samples* is $19-3=16$. A similar argument applies to the number of degrees of freedom for examining variation of counts within leaves.

In this kind of nested analysis, the mean-square at any level estimates the variance at that level plus those for all lower levels. Tests of significance must therefore be made by comparing a mean-square with that next below it. We see that the mean-square for between leaves within samples is not significantly greater than that for between counts within leaves (for 16/180 degrees of freedom, $F=1.29$ and p is greater than 0.1) but that the mean-square for between samples shows a highly significant increase (for 3/16 degrees of freedom, $F=5.30$ and p is equal to 0.01).

The trial indicates, then, that the greatest efficiency in estimating the mean chloroplast number in a particular light intensity would be obtained by the use of a large number of plant samples per light intensity and counting the chloroplasts in one cell or one leaf of each. In practice, final decisions on the experimental design would take into account the relative times involved in replication at the several levels of sampling.

Problems

10–1 Re-examine the data of Problem 9–4, taking into account that the determinations were carried out in 12 equal batches, each batch containing one sample (taken at random) from each site. Repeat the analysis of variance, this time extracting the sum-of-squares for batches, treating batches as blocks.

How do the results of this analysis affect the conclusions concerning sites? What do the results tell us about the reliability of the potassium determinations?

10–2 Because of the relatively large variation between fish revealed by the preliminary test of Problem 9–1, a two-treatment experiment (with and without NaCN) was carried out with a randomized block design and with four fish as blocks.

The following results were obtained.

	Treatments	
	Without NaCN	*With NaCN*
Fish 1	1.54	1.10
	1.92	1.42
	2.26	1.04
Fish 2	1.52	1.31
	2.02	1.15
	1.91	1.51
Fish 3	1.00	0.79
	1.12	0.84
	1.13	0.86
Fish 4	1.58	1.24
	1.78	0.81
	1.52	1.32

Analyse the data by the methods of §§10.2 and 10.3. State, with 95% confidence limits, the effect of adding the NaCN to the medium.

10–3 As part of an investigation into factors controlling the morphogenesis of fern gametophytes, a 2 × 2 factorial experiment was set up in which suspensions of fern spores in nutrient medium were treated as follows:

(i) Control—nutrient medium only.
(ii) +80 p.p.m. (v/v) ethanol.
(iii) +20 p.p.m. (v/v) acetaldehyde.
(iv) +80 p.p.m. (v/v) ethanol + 20 p.p.m. (v/v) acetaldehyde.

Thirty days after sowing, 20 gametophytes from each treatment were taken at random and their lengths measured in arbitrary micrometer units. The results were as follows:

		Ethanol		
		−	+	
Acetaldehyde −	Σx	2255	2994	(5249)
	n	20	20	
	$\bar{x}$	112.75	149.70	
	Σx^2	269 159	453 356	
Acetaldehyde +		731	1233	(1964)
		20	20	
		36.55	61.65	
		30 269	77 543	
		(2986)	(4227)	(7213)

Complete the analysis of variance and, if you think it appropriate, derive quantitative statements about the effect of 80 p.p.m. ethanol and the effect of 20 p.p.m. acetaldehyde on the development of the gametophytes.

10–4 At the same time the lengths of the gametophytes referred to above were taken, their total cell number was also recorded. The results, after transformation to $\sqrt{x}$, were as follows:

Acetaldehyde		*Ethanol* −	*Ethanol* +	
−	Σx	76.281	62.014	(138.295)
	n	20	62.014	
	$\bar{x}$	3.814	3.101	
	Σx^2	295	197	
+		38.034	47.992	(86.026)
		20	20	
		1.902	2.400	
		76	119	
		(114.315)	(110.006)	(224.321)

Complete the analysis of variance. What conclusions do you draw?

Solutions to Problems

Chapter 1

1–1

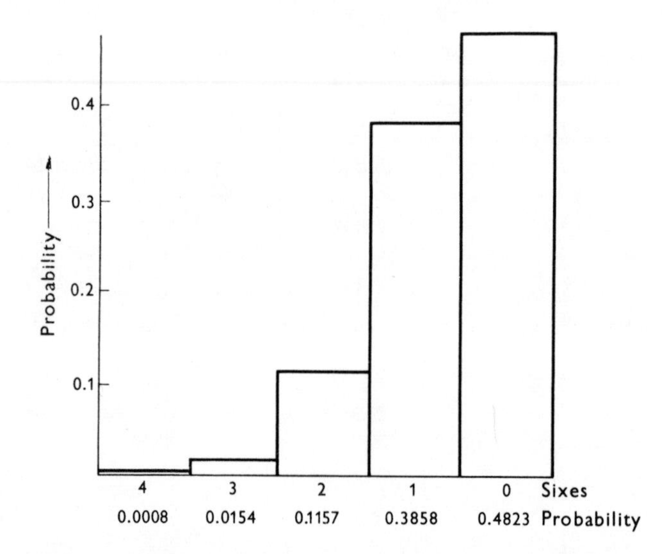

1–2 The probability of getting a particular question correct at random is 0.2 (i.e. $p=0.2$). The number of questions in the paper is six (i.e. $n=6$). So the distribution of probability of getting 6 questions correct, 5 correct, 4 correct, 3 correct, etc., is given by:

$(p+q)^n$ where $p=0.2$, $q=0.8$ and $n=6$

The seven terms from this expansion are:

No. correct	6	5	4	3	2	1	0
Probability	0.0001	0.0015	0.0154	0.0819	0.2458	0.3932	0.2621

In order to score 50% or more, 3, 4, 5 or 6 questions must be answered correctly. The sum of the three probabilities corresponding to these kinds of result is 0.0989 which is rather less than 1 : 10. A candidate therefore has about 10% chance of passing the paper if he selects his answers at random. This may seem a small probability from the point of view of the individual candidate but from the point of view of the examiners it means that out of a class of 100, on the average, 10 candidates could pass the examination knowing nothing!

Chapter 2

2–1 Two methods of calculation will be illustrated:

(a) When we have only a modest number of individual measurements, especially when these have been made to a fair degree of accuracy, and we have the use of a calculating machine, we usually calculate the sum (Σx) and the sum of the squares (Σx^2) directly from the individual measurements. On most machines both Σx and Σx^2 may be accumulated simultaneously, but in order to provide a check on possible human error it is advisable to first compute Σx alone and then to compute Σx together with Σx^2. If the two computed values of Σx agree then the computations are accepted as correct.

For the data of Problem 2–1:

$$\text{Sample mean, } \bar{x} = \frac{414.9}{30} = \underline{13.83}$$

$$\text{Standard deviation, } s = \sqrt{\left(\frac{5802.47 - 414.9^2/30}{29}\right)}$$

$$= \sqrt{\frac{64.40}{29}} = \underline{1.49}$$

$$\text{Standard error or the mean} = \frac{1.49}{\sqrt{30}} = \underline{0.27}$$

$$95\% \text{ confidence limits are } 13.83 \pm 1.96 \times 0.27$$

$$= \underline{13.83 + 0.53} \quad \text{or} \quad \underline{14.36 \text{ to } 13.30}$$

The conclusion drawn is that our best estimate of the mean of the population from which our sample of 30 was drawn is 13.83, and on consideration of the number and variation of the individual measurements we estimate that there is a probability of 0.95 that the true mean of the whole population lies between 14.36 and 13.30.

(b). When we have a very large number of individual measurements, especially when these are not made to a high degree of accuracy and/or we have not the use of a calculating machine, we often employ a rather different method of computing Σx and

Σx^2. We first display the data in terms of the frequencies with which the individual measurements fall into certain size classes. The size classes may result directly from the approximations made at the time of recording *or* they may be introduced as a means of simplification. For this exercise, we will use size classes of 1.0 and will obtain the necessary grouping by rounding-off the measurements to the nearest unit. In dealing with the two values 13.5 and 12.5, we adopt the convention of rounding-off such values alternately up to the next higher unit and down to the next lower unit.

Class (c)	Frequency (f)	Class × frequency ($c \times f$)	Class² (c^2)	Class² × frequency ($c^2 \times f$)
10	1	10	100	100
11	1	11	121	121
12	3	36	144	432
13	7	91	169	1183
14	8	112	196	1568
15	6	90	225	1350
16	3	48	256	768
17	1	17	289	289
	$n = 30$	$\Sigma x = 415$		$\Sigma x^2 = 5811$

Note that instead of adding in each value of x and x^2 separately we count the number of measurements which fall into each size class, calculate the contributions made by the whole class and then add in these.

(N.B. As might be expected, the rounding-off of the individual measurements has introduced a small error. With the large samples for which this method would normally apply this error becomes negligible.)

2–2 We evaluate $d = $(deviation of mean)/(standard error of mean) $= 0.7/0.27 = \underline{2.593}$. We see that this value approximates to that of 2.576 which corresponds to a probability of 0.01. There is thus a probability of approximately 0.01 that the sample mean deviates from the true mean by 0.7 or more.

2–3 The 95% confidence limits are to be ± 1.0 mm. Thus, the standard error of the mean $= 1.0/1.96$. But the standard error of the mean $= s\sqrt{n}$. So $1.0/1.96 = 5.0/\sqrt{n}$, where n is number of measurements to give the required confidence limits. Thus $\sqrt{n} = 9.8$ and $n = 96.04$. Thus, using the standard deviation estimated from the first 36 measurements it is calculated that a sample of 96 measurements would be required to provide an estimate of the population mean of the required accuracy. Since 36 measurements have already been made, *60 more measurements* are now required.

2–4 95% confidence limits are $\bar{x} \pm 1.960 \times s/\sqrt{25}$. If the total number of determinations needed to give 99% limits of ± 2.0 mg l^{-1} is n then 99% confidence limits are $\bar{x} \pm 2.576 \times s/\sqrt{n}$. So $1.960s/\sqrt{25} = 2.576s/\sqrt{n}$, and $n = 43.18$, i.e. 43 to the nearest integer. Since 25 determinations have already been made, 18 further determinations are needed.

Chapter 3

3–1 (Using the statistic d)

$\Sigma x_1 = 414.9$ $\qquad$ $\Sigma x_2 = 388.8$

$\Sigma x_1^2 = 5802.47$ $\qquad$ $\Sigma x_2^2 = 5096.27$

$\bar{x}_1 = 13.83$ $\qquad$ $\bar{x}_2 = 12.96$

$$s_1^2 = \frac{5802.47 - 172\,142.01/30}{29} = 2.22 \qquad s_2^2 = \frac{5096.27 - 151\,165.44/30}{29} = 1.98$$

$$d = \frac{13.83 - 12.96}{\sqrt{\left(\dfrac{2.22 + 1.98}{30}\right)}} = \underline{2.325}$$

which corresponds to a probability of 0.02.
We conclude that the difference between the two samples means is significant.

3–2 $n = 6$; $\Sigma x = 309.9$; $\bar{x} = 51.65$; $\Sigma x^2 = 16\,533.83$

$$\text{Standard deviation, } s = \sqrt{\left(\frac{16\,533.83 - 309.9^2/6}{5}\right)}$$

Standard error of the mean

$$= \frac{s}{\sqrt{n}} = \sqrt{\left(\frac{16\,533.83 - 309.9^2/6}{5 \times 6}\right)} = 4.1932$$

$t_{0.05}$ for five degrees of freedom $= 2.571$

So 95% confidence limits are

$$51.65 \pm 2.571 \times 4.193$$
$$= \underline{51.65 \pm 10.78}$$

We conclude that the true mean of the total phosphate content for peat soils carrying vegetation of this particular type has a 95% probability of lying between 40.87 and 62.43 mg/100 g dry weight.

3–3 (Using the statistic t)

$$s^2 = \frac{1}{30+30-2}\left(5802.47 - \frac{172\,142.01}{30} + 5096.27 - \frac{151\,165.44}{30}\right)$$

$$= 2.100$$

$$\therefore\ s = \sqrt{2.100} = 1.449$$

and $t = \dfrac{13.83 - 12.96}{1.449 \times \sqrt{(2/30)}} = \underline{2.325}$ with 58 degrees of freedom

which corresponds to a probability of 0.02.

3–4 A: $n_A = 14$; $\Sigma x_A = 1210$; $\bar{x}_A = 86.43$; $\Sigma x_A{}^2 = 104\,590$

$$s_A{}^2 = \frac{104\,590 - 1210^2/14}{13} = 0.879$$

B: $n_B = 18$; $\Sigma x_B = 1530$; $\bar{x}_B = 85.00$; $\Sigma x_B{}^2 = 130\,074$

$$s_B{}^2 = \frac{130\,074 - 1530^2/18}{17} = 1.41$$

$\therefore\ t = \dfrac{86.43 - 85.00}{\sqrt{(0.879/14 + 1.41/18)}} = \underline{3.80}$ with 30 degrees of freedom

which corresponds to a probability less than 0.001.
The difference between samples means is therefore significant.

3–5 Example A: $n_1 = 5$; $\Sigma x_1 = 125$; $\bar{x}_1 = 25$; $\Sigma x_1{}^2 = 3125.10$
$n_2 = 5$; $\Sigma x_2 = 150$; $\bar{x}_2 = 30$; $\Sigma x_2{}^2 = 4500.10$

$$s^2 = \frac{1}{5+5-2}\left(3125.10 - \frac{125^2}{5} + 4500.10 - \frac{150^2}{5}\right) = 0.025$$

$\therefore\ t = \dfrac{5.0}{\sqrt{[0.025(\frac{1}{5} + \frac{1}{5})]}} = \underline{50.0}$ with eight degrees of freedom

which greatly exceeds the tabulated value of 5.041 corresponding to $p = 0.001$.

Example B: $n_1 = 5$; $\Sigma x_1 = 125$; $\bar{x}_1 = 25$; $\Sigma x_1{}^2 = 3191.22$
$n_2 = 5$; $\Sigma x_2 = 150$; $\bar{x}_2 = 30$; $\Sigma x_2{}^2 = 4620.54$

$$s^2 = \frac{1}{5+5-2}\left(3191.22 - \frac{125^2}{5} + 4620.54 - \frac{150^2}{5}\right) = 23.345$$

$\therefore\ t = \dfrac{5.0}{\sqrt{[23.345(\frac{1}{5} + \frac{1}{5})]}} = \underline{1.636}$ with eight degrees of freedom

which corresponds to a probability of between 0.1 and 0.2.

3–6 Mean difference $(\bar{z}) = 637/50 = 12.74$ cm

$$\text{Standard error of mean difference} = \sqrt{\left(\frac{14\,433 - 637^2/50}{50 \times 49}\right)} = 1.606$$

$\therefore$ $t = 12.74/1.606 = \underline{7.93}$ with 49 degrees of freedom which exceeds the tabulated value for a probability of 0.001 (~ 3.5). We conclude that the mean difference is very highly significant.

Chapter 4

4–1 (i)

Treatment 1

No. germinated per dish	5	4	3	2	1	0
Observed frequency	30	81	57	24	7	1
Class total	150	324	171	48	7	0

Total germinated $a_1 = 700$
Proportion germinated $p_1 = 700/1000 = 0.7$
(Germination percentage)$_1 = 70\%$

Treatment 2

No. germinated per dish	5	4	3	2	1	0
Observed frequency	9	47	83	57	4	0
Class totals	45	188	249	114	4	0

Total germinated $a_2 = 600$
Proportion germinated $p_2 = 600/1000 = 0.6$
(Germination percentage)$_2 = 60\%$

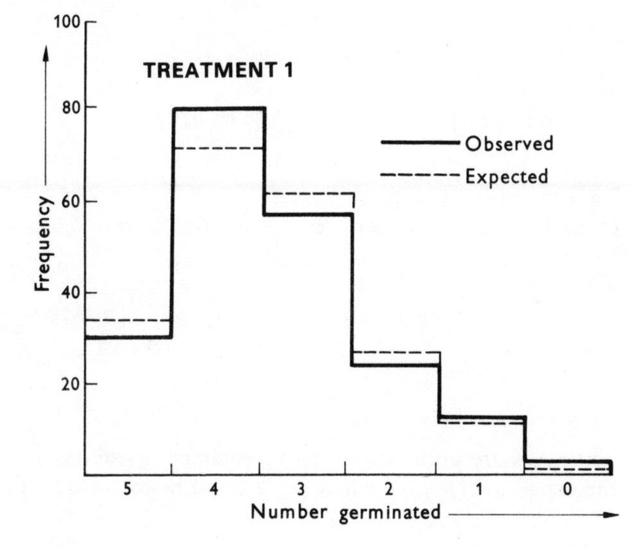

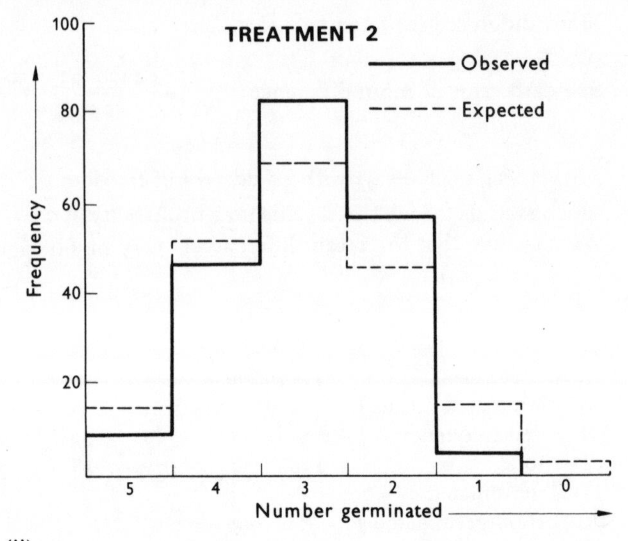

(ii)

No. germinated per dish	5	4	3	2	1	0
(Expected frequency)$_1$	33.61	72.03	61.74	26.46	5.67	0.49
(Expected frequency)$_2$	15.55	51.84	69.12	46.08	15.36	2.05

(iii)

Standard error of the estimate of $p_1 = \sqrt{[(0.7 \times 0.3)/1000]} = 0.0145$

Standard error of (germination percentage)$_1$ = $\underline{1.45\%}$

Standard error of the estimate of $p_2 = \sqrt{[(0.6 \times 0.4)/1000]} = 0.0155$

Standard error of (germination percentage)$_2$ = $\underline{1.55\%}$

(iv)

$70 \pm 1.96 \times 1.45\%$ $\quad$ $60 \pm 1.96 \times 1.55\%$

$= 70 \pm 2.84\%$ $\quad$ $= 60 \pm 3.04\%$

$= \underline{67.16 \text{ to } 72.84\%}$ $\quad$ $= \underline{56.96 \text{ to } 63.04\%}$

(v)

Total number of seeds set = 2000

Overall proportion germinated = (700 + 600)/2000 = 0.65

$$\text{Thus, } d = \frac{0.7 - 0.6}{\sqrt{\left[0.65\,(0.35)\left(\frac{1}{1000} + \frac{1}{1000}\right)\right]}} = \frac{0.1}{\sqrt{\left(0.2275 \times \frac{1}{500}\right)}}$$

$= 0.1/0.021\,33 = \underline{4.69}$

Therefore p is between 10^{-5} and 10^{-6}.

4–2 The necessary probabilities can be obtained by summing terms of the expansion $(p+q)^{10}$ where $p = 0.2$ and hence $q = 0.8$. The total probability for less than 2 (i.e. for 0 and 1) = 0.376 and the total probability for more than 2 (i.e. 3, 4, 5, 6, 7, etc.) = 0.322. The distribution of probability about the mean 2 is asymmetric as would be expected from a binomial expansion with $p \neq 0.5$. The lower probability of obtaining values more than 2 is compensated for by the possibility of obtaining a number deviating from the mean by more than 2, i.e. 5, 6, 7, etc. (In fact, the shopkeeper solved the boy's problem by serving him with a non-random sample of loose pastilles consisting of 10 blackcurrant ones!)

4-3 We may use the significance test for a difference between two proportions given in §4.4.

$p_1 = 0.875; \quad p_2 = 0.750$

$n_1 = 48; \quad n_2 = 48 \quad \therefore n = 96$

$$p = \frac{42 + 32}{96} = 0.8125$$

$$\therefore d = \frac{0.875 - 0.750}{\sqrt{\left[0.8125 \times 0.1875\left(\frac{1}{48} + \frac{1}{48}\right)\right]}} = \underline{1.57}$$

corresponding to a probability of about 0.12.
We conclude that the observed difference in hatchability was not significant.

4-4 Cover percentage $= 160/400 \times 100 = \underline{40\%}$
Standard error of estimated cover percentage $= 100\sqrt{(pq/n)}$

$$\text{or more conveniently} = \sqrt{(10^4 pq/n)} = \sqrt{\left(\frac{10^4 \times 0.4 \times 0.6}{400}\right)}$$

$$= \underline{2.45}$$

$\therefore$ 95% confidence limits of cover percentage are $40 \pm 1.96 \times 2.45\%$
$= \underline{44.80 \text{ to } 35.20\%}$

4-5 Cover percentage $= 120/400 \times 100 = \underline{30\%}$
We may use the significance test for a difference between two proportions given in §4.4.

$p_1 = 0.4; \quad p_2 = 0.3$

$n_1 = 400; \quad n_2 = 400 \quad \therefore n = 800$

$$p = \frac{160 + 120}{800} = 0.35$$

$$\therefore d = \frac{0.4 - 0.3}{\sqrt{\left[0.35 \times 0.65\left(\frac{1}{400} + \frac{1}{400}\right)\right]}} = \underline{2.965}$$

corresponding to a probability of about 0.002.

We conclude that there has been a highly significant fall in cover percentage of Daisies. (N.B. This may or may not have anything to do with the use of selective herbicide!)
(N.B. For an alternative method, suitable for Problems 4–3 and 4–5, see §7.2)

Chapter 5

5–1 (i) $N=100$; $\Sigma x=360$, thus $\bar{x}=360/100=\underline{3.60}$
(ii) Variance of 100 counts $=3.60$
Standard deviation of 100 counts $=\sqrt{3.60}$
Standard error of the mean of 100 counts $=\sqrt{(3.6/100)}=\underline{0.19}$
(iii) 95% confidence limits of the mean $=3.60\pm1.96\times0.19$
$=3.60\pm0.37=\underline{3.23 \text{ to } 3.97}$

5–2 The buns may be regarded as equal and random samples of a randomly distributed population of sultanas. We have to find Nm where $N=400$ and m is the mean number of sultanas per bun to give a probability of a bun taken at random of having no sultana of 0.001. In general, the probability of a sample containing no individual is e^{-m} so we put $e^{-m}=0.001$, thus $1/e^{m}=1/1000$ and $e^{m}=1000$. Taking logarithms of both sides, $m \log e=3.0$, thus $m=3.0/\log e=3.0/0.434\,29=6.908$. Thus $Nm=400\times6.980=2763.2$; i.e. $\underline{2763}$ to the nearest sultana.

5–3 The seed packets may be regarded as equal and random samples of the weed seed population. We are given that $m=1.5$ and have to find the total probability of from 4 to ∞ seeds occurring. We thus need the sum of the terms of the Poisson series $e^{-1.5}$ from the fifth term upwards. These terms are as follows: fours, 0.047 07; fives, 0.014 12; sixes, 0.003 53; sevens, 0.000 76, eights, 0.000 14 and nines, 0.000 02. The sum is thus $\underline{0.065\,64}$ which is equivalent to about 1:15.

5–4 Suppose that on each occasion he counts N samples containing a mean of m organisms, thus giving a total of Nm. The standard error of the mean of N counts $=\sqrt{(m/N)}$ and the 95% confidence limits of this mean are $m\pm1.96\sqrt{(m/N)}$. As the 95% limits are to be equal to 5%, $1.96\sqrt{(m/N)}=5m/100$. Squaring both sides and rearranging we have: $Nm=1.96^2\times400=1536.64$ giving a reasonable round number of $\underline{1500}$, or more accurately $\underline{1550}$.
(N.B. It is the total number of individuals counted which is important, not the number of samples.)

5–5 (i) $N=250$; $\Sigma x=500$; $\therefore\ \bar{x}=500/250=\underline{2.0}$

(ii)

No. of individuals	Probability	Expected frequency
0	0.135 34	33.83
1	0.270 67	67.67
2	0.270 67	67.67
3	0.180 45	45.11
4	0.090 22	22.56
5	0.036 09	9.02
6	0.012 03	3.01
7	0.003 44	0.86
8	0.000 86	0.21
9	0.000 19	0.05
10	0.000 04	0.01

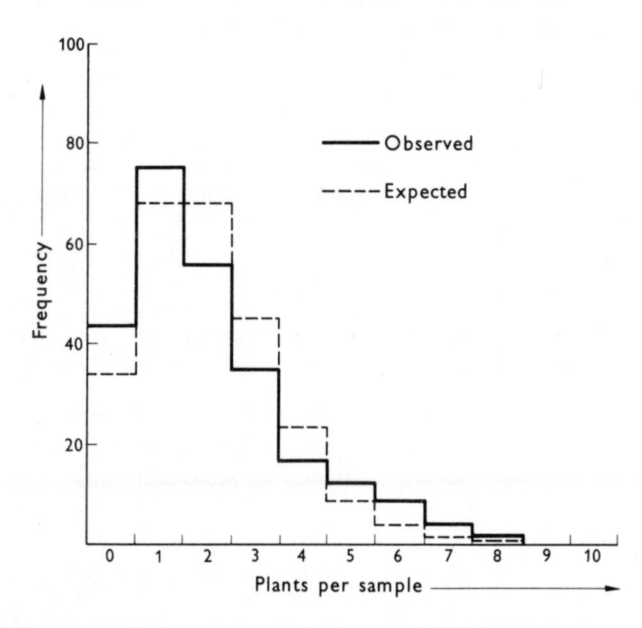

The observed frequencies of the smallest and largest size classes are greater than expected and the observed frequencies of the middle size classes are smaller.

(iii) When data are in the form of a frequency statement, Σx^2 is most conveniently calculated by evaluating the contribution of each size class and totalling these contributions.

No. of organisms per sample	0	1	2	3	4	5	6	7	8	9	10
(No. of organisms)^2 per sample	0	1	4	9	16	25	36	49	64	81	100
Frequency	43	75	56	34	17	12	9	3	1	0	0

We prepare the above table and accumulate on the calculating machine the sums of the products, $(0 \times 43) + (1 \times 75) + (4 \times 56) + (9 \times 34) + (16 \times 17) + \text{etc.}$ $\Sigma x^2 = 1712$.

$$s^2 = \frac{1712 - 250\,000/250}{249} = 712/249 = \underline{2.86}$$

This is to be compared with the expected variance of $\underline{2.0}$. As the observed variance is markedly greater than that expected on the basis of random distribution of individuals, we conclude that the individuals are non-randomly distributed and tend to be aggregated into groups.

Chapter 6

6–1

	Red-eyed	White-eyed	(Total)
Observed	768	818	(1586)
Expected	793	793	(1586)
O – E	– 25	+ 25	(0)
(O – E)^2/E	625/793	625/793	

$\Sigma\chi^2 = 1250/793 = \underline{1.576}$ with one degree of freedom

This corresponds to a probability of about 0.2. The deviation from expectation is therefore not significant.

6–2

	Fully-winged	Vestigial-winged	(Total)
Observed	2492	664	(3156)
Expected	2367	789	(3156)
O – E	+ 125	– 125	(0)
(O – E)^2/E	15 652/2367	15 652/789	

$\Sigma\chi^2 = 6.60 + 19.80 = \underline{26.4}$ with one degree of freedom

This corresponds to a probability of less than 0.001. The deviation from expectation is therefore highly significant. The cause of the deviation was sought and it was found that the numbers of vestigial-winged flies counted were markedly below expectation for certain culture bottles only. It was found that in these the food material had become unusually sticky and had apparently trapped many of the relatively less active vestigial-winged flies. Such trapped flies were not included in the counts.

6–3

Treatment 1							
No. germinated per dish	5	4	3	2	1	0	(1+0)
Observed frequency	30	81	57	24	7	1	(8)
Expected frequency	33.61	72.03	61.74	26.46	5.67	0.49	(6.16)
O − E	−3.61	+8.97	−4.74	−2.46	—	—	(+1.84)
$(O-E)^2/E$	0.388	1.117	0.364	0.229			(0.550)

$\Sigma\chi^2 = \underline{2.648}$ with three degrees of freedom

(N.B. Loss of additional degree of freedom because p estimated from the data.)
This corresponds to a probability of about 0.5, so the deviations from expectation are non-significant.

Treatment 2							
No. germinated per dish	5	4	3	2	1	0	(1+0)
Observed frequency	9	47	83	57	4	0	(4)
Expected frequency	15.55	51.84	69.12	46.08	15.36	2.05	(17.41)
O − E	−6.55	−4.84	+13.88	+10.92	—	—	−13.41
$(O-E)^2/E$	2.759	0.452	2.787	2.588			10.329

$\Sigma\chi^2 = \underline{18.915}$ with three degrees of freedom

This corresponds to a probability of less than 0.001, so the deviations from expectation are highly significant. The expected frequencies were calculated on the assumption of a fit to a binomial expansion, which in turn assumes that the seeds germinate independently of one another. We note that the observed frequencies of fives, fours, ones and zeros are lower than expected and that those of threes and twos are higher. This suggests that the germinating seeds tend to inhibit the germination of other seeds in the same dish (perhaps by liberating an inhibitor). Such a suggestion is consistent with both the lower germination percentage with treatment 2 and the fact that this treatment involves less complete leaching of the seeds. (How would you test this new hypothesis?)

6–4

No. of organisms/ sample	0	1	2	3	4		
Observed frequency	43	75	56	34	17		
Expected frequency	33.83	67.67	67.67	45.11	22.56		
O − E	9.17	7.33	−11.67	−11.11	−5.56		
(O − E)²/E	2.486	0.794	2.013	2.736	1.370		
No. of organisms/ sample	**5**	**6**	**7**	**8**	**9**	**10**	**(5–10)**
Observed frequency	12	9	3	1	0	0	(25)
Expected frequency	9.02	3.01	0.86	0.21	0.05	0.01	(13.16)
O − E							(11.84)
(O − E)²/E							(10.65)

$\Sigma\chi^2 = \underline{20.05}$ with four degrees of freedom

This corresponds to a probability of less than 0.001, so the deviations from expectation are highly significant. This result confirms that of the (variance)/(mean) ratio test in showing that the deviation from a random distribution is highly significant.

Chapter 7

7–1

		Clover +	*Clover* −	
Worm casts	+	90 (*a*)	25 (*b*)	(115)
	−	170 (*c*)	115 (*d*)	(285)
		(260)	(140)	(400)

$$x^2 = \frac{400(90 \times 115 - 25 \times 170 - 200)^2}{115 \times 285 \times 260 \times 140}$$

$$= \frac{400(10\,350 - 4250 - 200)^2}{1193 \times 10^6}$$

$$= \frac{4 \times 3481}{1193} = \underline{11.67} \text{ with one degree of freedom}$$

This corresponds with a probability of less than 0.001. Highly significant association is therefore indicated. Since *ad* exceeds *bc*, the association is positive.

7–2

	4-day	6-day	(Total)
Hatched	42	36	(78)
Failed to hatch	6	12	(18)
(Total)	(48)	(48)	(96)

$$\chi^2 = \frac{96(42 \times 12 - 36 \times 6 - 48)^2}{78 \times 18 \times 48 \times 48} = \underline{1.709} \text{ with one degree of freedom}$$

This corresponds to a probability greater than 0.1

7-3 Set up a 6×2 table and calculate all expected values from the marginal totals, thus:

	<30	30–39	40–49	50–59	60–69	⩾70	(Total)
Group A							
Observed	3	8	32	18	12	5	(78)
Expected	4.98	11.06	29.87	16.60	11.06	4.43	
Group B							
Observed	6	12	22	12	8	3	(63)
Expected	4.02	8.94	24.13	13.40	8.94	3.57	
(Total)	(9)	(20)	(54)	(30)	(20)	(8)	(141)

We note that four of the expected frequencies are less than 5, but since none approaches unity we decide that the analysis may be completed. (Alternatively, the observed frequencies for both <30 and ⩾70 could be added to those in the adjacent columns to yield a 4×2 table.)

Calculate χ^2 in the form $\Sigma(O^2/E) - n$ thus:

$$\chi^2 = 1.81 + 5.79 + 34.28 + 19.52 + 13.02 + 5.64 + 8.96 + 16.11 + 20.06 + 10.75 + 7.16 + 2.52 - 141$$
$$= \underline{4.62} \text{ with five degrees of freedom}$$

This corresponds to a probability of about 0.5, so the results do not show significant heterogeneity.

7-4 We now have a 2×4 table, thus:

Batch	Wild type	Mutant	(Total)
1	60	26	(86)
2	75	33	(108)
3	81	18	(99)
5	54	21	(75)
(Total)	(270)	(98)	(368)

$$\chi^2 \text{ value for overall segregation} = \frac{(270-276)^2}{276} + \frac{(98-92)^2}{92}$$
$$= \underline{0.522} \text{ with one degree of freedom}$$

Total χ^2 value (sum of individual χ^2 values for four batches) $= 1.256 + 1.778 + 2.455 + 0.360 = \underline{5.849}$ with four degrees of freedom

Hence calculate χ^2 for heterogeneity by subtraction and display the results:

Sources of variation	Degrees of freedom	x^2	Probability
Deviation from 3:1 ratio	1	0.522	more than 0.3
Heterogeneity	3	5.327	more than 0.1
(Total)	(4)	(5.849)	—

We conclude that the experimental data are homogeneous and do not deviate significantly from the expected ratio of 3:1.

7–5 Set up a 2×4 table and insert expected values, thus:

	Male	(Expected)	Female	(Total)
Control	34	(32.5)	31	(65)
pH 7.0	22	(19.0)	16	(38)
pH 4.0	9	(15.0)	21	(30)
pH 9.2	21	(14.0)	7	(28)
(Total)	(86)	(80.5)	(75)	(161)

χ^2 value for overall fit to 1:1 ratio $= \frac{5.5^2}{80.5} \times 2 = \underline{0.752}$

Total χ^2 value (sum of individual χ^2 values for four treatments)

$$= \frac{1.5^2}{32.5} \times 2 + \frac{3.0^2}{19.0} \times 2 + \frac{6.0^2}{15.0} \times 2 + \frac{7.0^2}{14.0} \times 2$$

$$= 0.138 + 0.947 + 4.800 + 7.000$$

$$= \underline{12.885}$$

Hence calculate χ^2 for heterogeneity by subtraction and display the results, thus:

Sources of variation	Degrees of freedom	χ^2	Probability
Deviation from 1:1 ratio	1	0.752	between 0.1 and 0.5
Heterogeneity	3	12.133	less than 0.01
(Total)	(4)	(12.885)	—

Having established highly-significant heterogeneity, re-examine the table and note that the frequencies for both pH 4.0 and pH 9.2 deviate markedly from expectation with individual χ^2 values of 4.80 and 7.00, corresponding to probabilities of less than 0.05 and less than 0.01, respectively. Also note that the frequencies for control and for pH 7.0 deviate little with χ^2 values of 0.138 and 0.947 respectively, both corresponding to probabilities greater than 0.1. Further note that the deviations for pH 4.0 and pH 9.2 are in opposite directions. Hence we conclude that vaginal pH does affect sex ratio in mice, acidity increasing the proportion of females and alkalinity the proportion of males.

7–6

	G	g	(Total)
H	123 (a)	27 (b)	(150)
h	30 (c)	21 (d)	(51)
(Total)	(153)	(48)	(201)

For G–g, $\chi^2 = \frac{1}{3n}[(a+c)-3(b+d)]^2$

$= \frac{1}{603}(153-144)^2 = \underline{0.134}$ with one degree of freedom

corresponding to a probability p between 0.8 and 0.7.

For H–h, $\chi^2 = \frac{1}{3n}[(a+b)-3(c+d)]^2$

$= \frac{1}{603}(150-153)^2 = \underline{0.015}$ with one degree of freedom

corresponding to a probability p between 0.95 and 0.90.

For linkage, $\chi^2 = \frac{1}{9n}[(a+9d)-3(b+c)]^2$

$= \frac{1}{1809}(123+189-3\times 57)^2$

$= \underline{10.99}$ with one degree of freedom

corresponding to a probability p of less than 0.001.

We conclude that there is no significant deviation from the hypothesis which suggests a 3:1 segregation for G:g and H:h but that there is highly significant linkage.

Chapter 8

8–1

$$t = \frac{1.08 - 0.06}{s\sqrt{\left(\frac{1}{15} + \frac{1}{25}\right)}} \text{ where } s \text{ is given by:}$$

$$s^2 = \frac{1}{25 + 25 - 2}(14 \times 0.019 + 24 \times 0.0158)$$

$$= \frac{1}{38}(0.266 + 0.379)$$

$$= 0.017$$

Thus $s = 0.13$ and $t = \frac{0.12}{0.13 \times 0.327} = \underline{2.82}$

This value is rather more than the tabulated value of t for 38 degrees of freedom and $p = 0.01$ (~ 2.75), so we conclude that the presence of *T. repens* is associated with a highly significantly greater quantity of total nitrogen in the surface soil than its absence.

8–2

$$\Sigma x = 38.37 \qquad \Sigma y = 7.491$$

$$\Sigma x^2 = 64.344 \qquad \Sigma y^2 = 4.328$$

$$\frac{(\Sigma x)^2}{n} = 61.344 \qquad \frac{(\Sigma y)^2}{n} = 2.338$$

$$\Sigma x^2 - \frac{(\Sigma x^2)}{n} = 3.000 \qquad \Sigma y^2 - \frac{(\Sigma y)^2}{n} = 1.990$$

$$n = 24$$

$$\Sigma xy = 10.295$$

$$\frac{(\Sigma x)(\Sigma y)}{n} = 11.976$$

$$\Sigma xy - \frac{(\Sigma x)(\Sigma y)}{n} = -1.681$$

(a) $$b = \frac{-1.681}{3.000} = -0.560; \quad \bar{x} = \frac{38.37}{24} = 1.599;$$

$$\bar{y} = \frac{7.491}{24} = 0.312$$

$$a = 0.312 + (0.560 \times 1.599)$$

$$= 0.312 + 0.895$$

$$= 1.207$$

So regression equation becomes $\underline{y = 1.207 - 0.560x}$

(b) To test significance of departure of b from zero, first calculate the residual variance:

$$s^2 = \frac{1}{22}\left(1.990 - \frac{1.681^2}{3.000}\right)$$

$$= \frac{1}{22}(1.990 - 0.942)$$

$$= 0.0476$$

Then $s = 0.2182$ and standard error of estimate of $b = 0.2182/\sqrt{3.000} = 0.1260$ and $t = 0.560/0.1260 = \underline{4.44}$.

This value exceeds the tabulated value of t for 22 degrees of freedom and $p = 0.001$ (3.792), so we conclude that the regression is highly significant.

(c) $r = \dfrac{-1.681}{\sqrt{(3.000 \times 1.990)}} = \dfrac{-1.681}{2.443} = -\underline{0.688}$

(d) $t = \dfrac{-0.688\sqrt{22}}{\sqrt{(1 - 0.688^2)}} = \dfrac{0.688 \times 4.690}{0.7257} = \underline{4.45}$

(N.B. The value of r calculated under (c) above just exceeds the tabulated value of r for 22 degrees of freedom and $p = 0.001$ (~ 0.63) and from this we would conclude the correlation to be highly significant. A similar conclusion is of course reached by entering the value of $t = 4.45$ in the table of t.)

8–3 $b = \dfrac{42.835}{5.0238} = 8.526$; $\bar{x} = \dfrac{40.92}{40} = 1.023$; $\bar{y} = \dfrac{249.36}{40} = 6.234$;

$a = 6.234 - (8.526 \times 1.023) = 2.488$

So $\underline{y = 8.526x - 2.488}$

Let $x = 0.5$ then $y = 1.775$. Let $x = 1.5$ then $y = 10.301$

$$s_R^2 = \frac{1}{38}\left(483.740 - \frac{42.835^2}{5.0238}\right) = 3.119 \quad \therefore s_R = 1.766$$

Standard error of $b = 1.766/\sqrt{5.0238} = 0.788$

and $t = 8.526/0.788 = \underline{10.82}$ with 38 degrees of freedom

This corresponds to a probability of less than 0.001.

Confidence limits of the estimate (large sample)

$$= yp \pm 1.96 \times 1.766$$
$$= yp \pm 3.46$$

8–4 First find d, the difference in ranking order for each sample, e.g. for both samples 9 and 17, $d = 2$; for sample 5, $d = 5$; for sample 15, $d = 3$, etc. Calculate $\Sigma d^2 = 216$ and then:

$$r_s = 1 - \frac{6 \times 216}{24(24^2 - 1)} = 1 - \frac{6 \times 216}{13\,800} = 1 - 0.094$$

$$= \underline{0.906}$$

As n is greater than 8, enter this value (0.906) into the table of r (product–moment correlation coefficient). This value exceeds the tabulated value for 22 degrees of freedom and $p = 0.001$ (~ 6.3), so we conclude that the two ordinations are highly significantly correlated.

Chapter 9

9–1 First calculate the necessary sums and sums-of-squares, thus:

$T_1 = 6.91$; $T_2 = 5.97$; $T_3 = 5.05$; $T_4 = 3.67$;
$GT = 21.60$; $\Sigma x_1^2 = 16.04$; $\Sigma x_2^2 = 11.89$; $\Sigma x_3^2 = 8.60$;
$\Sigma x_4^2 = 4.52$; $\Sigma x^2 = 41.05$

Then,

$$C = \frac{21.60^2}{12} = 38.88$$

$$SS = 41.05 - 38.88 = 2.17$$

$$SST = \frac{6.91^2 + 5.97^2 + 5.05^2 + 3.67^2}{3} - 38.88 = 1.91$$

$$SSE = 2.17 - 1.91 = 0.26$$

Sources of variation	Sums-of-squares	Degrees of freedom	Mean-squares	F
Between fish	1.91	3	0.637	19.6
Between replicates	0.26	8	0.0325	—
(Total)	(2.17)	(11)	—	—

The between-fish variance is about 20 times that between replicates for the same fish. The value for F with 3 (numerator) and 8 (denominator) degrees of freedom exceeds the tabulated value for a probability of 0.001.

9–2 First calculate the grand total (GT) and total sums-of-squares, thus:

$GT = 338 + 1031 + 1097 + 1034 + 1085 + 397 + 684 = 5666$
$\Sigma x^2 = 17\,096 + 53\,297 + 54\,305 + 50\,050 + 52\,783 + 19\,965 + 32\,740 = 280\,236$

Then,

$$C = \frac{5666^2}{120} = 267\,530$$

$$SS = 280\,236 - C = 12\,706$$

$$SST = \frac{338^2}{7} + \frac{1031^2}{21} + \frac{1097^2}{23} + \frac{1034^2}{23} + \frac{1085^2}{23} + \frac{397^2}{8} + \frac{684^2}{15} - C$$

$$= 290$$

$$SSE = 12\,706 - 290 = 12\,416$$

Sources of variation	Sums-of-squares	Degrees of freedom	Mean-squares	F
Past experience and performance	290	6	48.33	0.44
Within groups	12 416	113	109.87	—
(Total)	(12 706)	(119)	—	—

The value for F with 6 (numerator) and 113 (denominator) degrees of freedom corresponds to a probability of more than 0.75. We retain the null hypothesis of independence. The variation between the groups, when compared with the variation within the groups, is too small for us to discard the hypothesis that the true means of the seven groups are the same.

9–3

$$C = \frac{190.874^2}{50} = 728.658$$

$$SS = 730.195 - C = 1.537$$

$$SST = 729.242 - C = 0.584$$

$$SSE = 1.537 - 0.584 = 0.953$$

Sources of variation	Sums-of-squares	Degrees of freedom	Mean-squares	F
Treatments	0.584	4	0.146	6.9
Error	0.953	45	0.0212	—
(Total)	(1.537)	—	—	—

This F value is greater than the tabulated value for 4/45 degrees of freedom at $p = 0.001$ (5.37) so the differences between treatments are very highly significant.

(a) L.S.D. $= t_{0.05}\sqrt{(2s^2/n)}$ where $t_{0.05}$ is the tabulated value of t for $p = 0.05$ and 45 degrees of freedom

So, L.S.D. $= 2.14\sqrt{\left(\frac{2 \times 0.0212}{10}\right)} = \underline{0.139}$

The mean transformed value for red light (3.604) differs by more than the L.S.D. from the mean of each other treatment.
(b) L.S.R. $= Q_{0.05}\sqrt{(s^2/n)}$ where $Q_{0.05}$ is the tabulated value of Q for five treatments and 45 degrees of freedom. For $p=0.05$, $Q_{0.05}=4.02$.

$$\text{So, L.S.R.} = 4.02\sqrt{\left(\frac{0.0212}{10}\right)} = \underline{0.185}$$

The mean transformed value for red light (3.604) differs by more than the L.S.R. from the mean of each other treatment.
(N.B. Although here the two tests lead to the same conclusion, they will not always do so. It will be noted that the L.S.R. is substantially larger than the L.S.D.)
It could be argued that in this experiment it would be more appropriate to compare the mean for red light with the mean for all the other colours *taken together*. This can be done quite simply by partitioning the sum-of-squares for between treatments (with four degrees of freedom) into two components: one for a comparison between red and the other colours taken together, and the other for comparisons between the other colours.

Thus:

$$\Sigma(\log x)_{\text{red}} = 36.037 \quad \text{and} \quad \Sigma(\log x)_{\text{others}} = 154.837$$

$$SS_{\text{red/others}} = \frac{36.037^2}{10} + \frac{154.837^2}{40} - C = 729.229 - C = 0.571$$

$$SS_{\text{others}} = SST - SS_{\text{red/others}} = 0.584 - 0.571 = 0.013$$

Sources of variation	Sums-of-squares		Degrees of freedom	Mean-squares	F
Treatments:	0.584		4	0.146	6.9
red/others		0.571	1	0.571	26.6
others		0.013	3	0.0043	0.2
Error	0.953		45	0.0212	—
(Total)	(1.537)		(49)	—	—

The value of F with 1/45 degrees of freedom is 26.6 which corresponds to a p value of less than 0.001; that with 3/45 degrees of freedom is less than 1.0

9–4

$$C = \frac{297\,183\,1.21}{120} = 24\,765.26$$

$$SS = 25\,640.45 - C = 875.19$$

$$\text{SST} = \frac{305\,078.01}{12} - \text{C} = 657.91$$

$$\text{SSE} = 875.19 - 657.91 = 217.28$$

Sources of variation	Sums-of-squares	Degrees of freedom	Mean-squares	F
Between sites	657.91	9	73.1	37.0
Within sites (error)	217.28	110	1.975	—
(Total)	(875.19)	(119)	—	—

This value of F greatly exceeds the tabulated value for 9/110 degrees of freedom and $p = 0.001$ (~ 3.3). We therefore conclude that the between-site differences are very highly significant.

$$\text{Standard error of site mean} = \sqrt{\left(\frac{1.975}{12}\right)} = 0.4057$$

'Studentized ranges' for multiple-range test for 5% level of significance and 110 degrees of freedom:

Group size	2	3	4	5	6	7	8	9	10
Q	2.80	3.36	3.69	3.90	4.10	4.24	4.36	4.47	4.56
S.S.R.*	1.14	1.36	1.50	1.58	1.66	1.72	1.77	1.81	1.85

Using these ranges the site means are found to fall into the following groups: 10; 5 and 7; 8 and 9; 6, 4 and 3; 1 and 2.

9–5 Total sum-of-squares for $y = 1.990$
Regression sum-of-squares $= 0.942$
Residual sum-of-squares $= 1.990 - 0.942 = 1.048$

Sources of variation	Sums-of-squares	Degrees of freedom	Mean-squares	F
Regression	0.942	1	0.942	19.6
Residual	1.048	22	0.048	—
(Total)	(1.990)	(23)	—	—

This value of F exceeds the tabulated value of F for 1/22 degrees of freedom and $p = 0.001$ (14.4), so the regression is very highly significant.

* Shortest significant ranges, each calculated as 'Studentized range' × standard error of the mean (0.4057).

Chapter 10

10–1 The analysis requires the use of the batch totals which are included on the right-hand side of the table of results.

$$C = 24\,765.26$$
$$SS = 875.19$$
$$SST = 657.91$$
$$SSB = \frac{247\,871.57}{10} - C = 21.90$$
$$SSE = 875.19 - (675.91 + 21.90) = 195.38$$

Sources of variation	Sums-of-squares	Degrees of freedom	Mean-squares	*F*
Sites	657.91	9	73.1	37.4
Batches	21.90	11	1.99	1.02
Error	195.38	100	1.95	—
(Total)	(875.19)	(119)	—	—

The main conclusion to be drawn from the analysis remains the same: the level of significance of the *F* value for sites has remained more or less unchanged because of the small size of the sum-of-squares for batches. The variation in batches is shown to be small and no larger than would be expected from replicates analysed at the same time. There is therefore no evidence of any systematic error between batches.

10–2 First calculate sub-class, treatment, block and grand totals thus:

Fish	Treatments: Without NaCN	With NaCN	(Total)
1	5.72	3.56	(9.28)
2	5.45	3.97	(9.42)
3	3.25	2.49	(5.74)
4	4.88	3.37	(8.25)
(Total)	(19.30)	(13.39)	(32.69)

Then:

$$C = \frac{32.69^2}{24} = 44.5265$$

$$SS = 1.54^2 + 1.92^2 + 2.26^2 + 1.10^2 + \ldots + 1.32^2 - C$$
$$= 48.3431 - C = 3.8166$$

$$SST = \frac{19.30^2 + 13.39^2}{12} - C = 1.4553$$

$$SSB = \frac{9.28^2 + 9.42^2 + 5.74^2 + 8.25^2}{6} - C = 1.4510$$

$$SSE = 3.8166 - (1.4553 + 1.4510) = 0.9103$$

Sources of variation	Sums-of-squares	Degrees of freedom	Mean-squares	F
Treatments	1.4553	1	1.4553	30.38
Fish	1.4510	3	0.4837	10.10
Error	0.9103	19	0.0479	—
(Total)	(3.8166)	(23)	—	—

F with 1/19 degrees of freedom is equal to 30.38 and p is less than 0.001. F with 3/19 degrees of freedom is equal to 10.10 and p is less than 0.001. From this we conclude that the differences between treatments and between individual fish are highly significant.

Now examine for interaction:

$$SSQ = \frac{5.72^2 + 5.45^2 + 3.25^2 + \ldots + 3.37^2}{3} - C = 3.0699$$
$$SSE = 3.8166 - 3.0699 = 0.7467$$
$$SSTB = 3.0699 - (1.4553 + 1.4510) = 0.1636$$

Sources of variation	Sums-of-squares	Degrees of freedom	Mean-squares	F
Treatments	1.4553	1	1.4553	31.16
Fish	1.4510	3	0.4837	10.36
Interaction	0.1636	3	0.0545	1.167
Error	0.7467	16	0.0467	—
(Total)	(3.8166)	(23)	—	—

F for interaction, with 3/16 degrees of freedom is 1.167 corresponding to a value of p greater than 0.25.

As interaction is non-significant we can make a quantitative statement about the difference between treatments which is applicable to all blocks, i.e.

$$\text{Difference between treatment means} = \frac{19.30 - 13.39}{12} = 0.49$$

The 95% confidence limits of the difference between the two means are given by $t_{0.05}\sqrt{(2s^2/n)}$ where the value of t is for 16 degrees of freedom, s^2 is the error mean square for error in the second analysis and n is the number of replicates for each treatment.

$$95\% \text{ confidence limits of difference} = 2.120\sqrt{\left(\frac{2 \times 0.0467}{12}\right)}$$
$$= 0.19$$

Thus, we may state that the addition of the NaCN to the medium reduced uptake by 0.49 ± 0.19 μmol g^{-1} dry weight per 20 min period.

10–3

$$C = 650\,342$$
$$SS = 830\,327 - C = 179\,985$$
$$SST = 669\,593 - C = 19\,251$$
$$SSB = 785\,232 - C = 134\,890$$
$$SSQ = 805\,186 - C = 154\,844$$
$$SSE = SS - SSQ = 25\,141$$
$$SSTB = SSQ - (SST + SSB) = 703$$

Sources of variation	Sums-of-squares	Degrees of freedom	Mean-squares	F
Ethanol	19 251	1	19 251	58.16
Acetaldehyde	134 890	1	134 890	407.5
Interaction	703	1	703	2.124
Error	25 141	76	331	—
(Total)	(179 985)	(79)	—	—

F for interaction, with 1/76 degrees of freedom, is 2.124 and p is greater than 0.1. For the effects of ethanol and acetaldehyde treatments, F again with 1/76 degrees of freedom, is respectively 58.16 and 407.5, and p is less than 0.001.

We conclude that the level of interaction is non-significant but that the effects of ethanol and acetaldehyde are both very highly significant. We can therefore make quantitative statements about these effects.

Effect of ethanol

Difference in means = 31.0

$s^2 = 331$, so $s = 18.19$

$$\text{Standard error of difference} = s\sqrt{\left(\frac{1}{20}+\frac{1}{20}\right)} = 5.75$$

95% confidence limits of difference

$= 31.0 \pm t_{0.05} \times 5.75$ (with 76 degrees of freedom)

$= 31.0 \pm 2.0 \times 5.75$

$= 31.0 \pm 11.5$

Therefore 80 p.p.m. ethanol *increases* gametophyte length by 31.0 ± 11.5 units.

Effect of acetaldehyde

Difference in means = 82.1

95% limits of difference = 82.1 ± 11.5

Therefore 20 p.p.m. acetaldehyde *decreases* gametophyte length by 82.1 ± 11.5 units.

10–4

$$C = \frac{224.321^2}{80} = 629.00$$

$SS = 687 - C = 58.00$

$SSQ = 670.72 - C = 41.72$

$SSE = 58.00 - 41.72 = 16.28$

$SST = 629.23 - C = 0.23$

$SSB = 663.15 - C = 34.15$

$SSTB = 41.72 - (0.23 + 34.15) = 7.34$

Sources of variation	Sums-of-squares	Degrees of freedom	Mean-squares	F
Ethanol	0.23	1	0.23	—
Acetaldehyde	34.15	1	34.15	—
Interaction	7.34	1	7.34	34.27
Error	17.28	76	0.2142	—
(Total)	(56.0)	(79)	—	—

F for interaction, with 1/76 degrees of freedom, is equal to 34.27 giving a value of p less than 0.001.

We therefore conclude that interaction is very highly significant. Thus we set up a table of sub-class means, both observed and expected together with their difference.

		Ethanol −	Ethanol +	
Acetaldehyde −	Observed	3.814	3.101	(6.915)
	Expected	3.515	3.407	
	Difference	+0.303	−0.303	
Acetaldehyde +	Observed	1.902	2.400	(4.302)
	Expected	2.205	2.097	
	Difference	−0.303	+0.303	
		(5.716)	(5.501)	(11.217)

Both ethanol and acetaldehyde appear to reduce cell number, strictly $\sqrt{(\text{cell number})}$, the former slightly and the latter very considerably. (See column and row totals.) In combination, the two substances have less effect than would be expected from their activities when applied alone, i.e. they show ***negative interaction.*** This interaction is highly significant and is one of the most interesting aspects of the results. There is little point in trying to make quantitative statements about the effects of ethanol and acetaldehyde unless we restrict ourselves to their effects when applied alone. To do this we can extract the data for two separate experiments: ±ethanol, and ±acetaldehyde.

	Ethanol −	Ethanol +	
Σx	76.281	62.014	(138.295)
n	20	20	
$\bar{x}$	3.814	3.101	
Σx^2	295	197	(492)

$$C = \frac{138.295^2}{40} = 478.138$$

$$SS = 492 - C = 13.862$$

$$SST = 483.226 - C = 5.088$$

$$SSE = 13.862 - 5.088 = 8.774$$

Sources of variation	Sums-of-squares	Degrees of freedom	Mean-squares	F
±Ethanol	5.088	1	5.088	22.04
Error	8.774	38	0.2309	—
(Total)	(13.862)	(39)	—	—

F with 1/38 degrees of freedom is equal to 22.04 and p is less than p.001.

Difference between means = 0.713

$s^2 = 0.2309$, so $s = 0.4805$ and standard error of difference

$$= 0.4805\sqrt{\left(\frac{1}{20}+\frac{1}{20}\right)} = 0.152$$

95% confidence limits of difference

$$= 0.713 \pm t_{0.05} \times 0.152 \text{ (with 38 degrees of freedom)}$$
$$= 0.713 \pm 2.02 \times 0.152$$
$$= 0.713 \pm 0.307$$

Acetaldehyde

	−	+	
Σx	76.281	38.034	(114.315)
n	20	20	
$\bar{x}$	3.841	1.902	
Σx^2	295	76	(371)

$$C = \frac{114.315^2}{40} = 326.698$$

$$SS = 371 - C = 44.302$$
$$SST = 363.269 - C = 36.571$$
$$SSE = 44.302 - 36.571 = 7.731$$

Sources of variation	Sums-of-squares	Degrees of freedom	Mean-squares	F
± Acetaldehyde	36.571	1	36.571	179.79
Error	7.731	38	0.2034	—
(Total)	(44.302)	(39)	—	—

F with 1/38 degrees of freedom is equal to 179.79 and p is much less than 0.001.

Difference between means = 1.912

$s^2 = 0.2034$, so $s = 0.4510$ and standard error of difference

$$= 0.4510\sqrt{\left(\frac{1}{20}+\frac{1}{20}\right)} = 0.143$$

95% confidence limits of difference

$$= 1.912 \pm t_{0.05} \times 0.143 \text{ (with 38 degrees of freedom)}$$
$$= 1.912 \pm 2.02 \times 0.143$$
$$= 1.912 \pm 0.289$$

(N.B. As no doubt has been thrown on the equality of the variance between replicates within the four sub-classes, confidence limits to treatment effects could have been estimated using the mean square for error for the whole data, i.e. 0.2142, and $t_{0.05}$ with 76 degrees of freedom.)

Further Reading

BAILEY, N. T. (1959). *Statistical Methods in Biology*. English Universities Press, London.
BISHOP, O. N. (1966). *Statistics for Biology*, Longmans, London.
CAMPBELL, R. C. (1967). *Statistics for Biologists*, Cambridge University Press, London.
CLARKE, G. M. (1969). *Statistics and Experimental Design*, Edward Arnold, London.
HEATH, O. V. S. (1970). *Investigation by Experiment*, Studies in Biology No. 23, Edward Arnold, London.
MATHER, K. (1949). *Statistical Analysis in Biology*, Methuen, London.
MATHER, K. (1967). *The Elements of Biometry*, Methuen, London.
SNEDECOR, G. W. and COCHRAN, W. G. (1967). *Statistical Methods*, 6th edn., Iowa State University Press.
SOKAL, R. R. and ROHLF, F. J. (1969). *Biometry*, Freeman, London.

Statistical tables

FISHER, R. A. and YATES, F. (1963). *Statistical Tables for Biological, Agricultural and Medical Research*, Oliver & Boyd, Edinburgh.
LINDLEY, D. V. and MILLER, J. C. P. (1968). *Cambridge Elementary Statistical Tables*, Cambridge University Press, London.
PEARSON, E. S. and HARTLEY, H. O. (1962). *Biometrika Tables for Statisticians*, Vol. 1, Cambridge University Press, London.
ROHLF, F. J. and SOKAL, R. R. (1969). *Statistical Tables*, Freeman, London.

Selected Statistical Tables

Table I Student's ***t***. Values exceeded with probability *p*

d.f.	$p=0.1$	0.05	0.02	0.01	0.002	0.001
1	6.314	12.706	31.821	63.657	318.31	636.62
2	2.920	4.303	6.965	9.925	22.327	31.598
3	2.353	3.182	4.541	5.841	10.214	12.924
4	2.132	2.776	3.747	4.604	7.173	8.610
5	2.015	2.571	3.365	4.032	5.893	6.869
6	1.943	2.447	3.143	3.707	5.208	5.959
7	1.895	2.365	2.998	3.499	4.785	5.408
8	1.860	2.306	2.896	3.355	4.501	5.041
9	1.833	2.262	2.821	3.250	4.297	4.781
10	1.812	2.228	2.764	3.169	4.144	4.587
11	1.796	2.201	2.718	3.106	4.025	4.437
12	1.782	2.179	2.681	3.055	3.930	4.318
13	1.771	2.160	2.650	3.012	3.852	4.221
14	1.761	2.145	2.624	2.977	3.787	4.140
15	1.753	2.131	2.602	2.947	3.733	4.073
16	1.746	2.120	2.583	2.921	3.686	4.015
17	1.740	2.110	2.567	2.898	3.646	3.965
18	1.734	2.101	2.552	2.878	3.610	3.922
19	1.729	2.093	2.539	2.861	3.579	3.883
20	1.725	2.086	2.528	2.845	3.552	3.850
21	1.721	2.080	2.518	2.831	3.527	3.819
22	1.717	2.074	2.508	2.819	3.505	3.792
23	1.714	2.069	2.500	2.807	3.485	3.767
24	1.711	2.064	2.492	2.797	3.467	3.745
25	1.708	2.060	2.485	2.787	3.450	3.725
26	1.706	2.056	2.479	2.779	3.435	3.707
27	1.703	2.052	2.473	2.771	3.421	3.690
28	1.701	2.048	2.467	2.763	3.408	3.674
29	1.699	2.045	2.462	2.756	3.396	3.659
30	1.697	2.042	2.457	2.750	3.385	3.646
40	1.684	2.021	2.423	2.704	3.307	3.551
60	1.671	2.000	2.390	2.660	3.232	3.460
120	1.658	1.980	2.358	2.617	3.160	3.373
∞	1.645	1.960	2.326	2.576	3.090	3.291

The last row of the table (∞) gives values of d, the 'standardized normal deviate'.

Table II χ^2

d.f.	p = **0.900**	**0.500**	**0.100**	**0.050**	**0.010**	**0.001**
1	0.016	0.455	2.71	3.84	6.63	10.83
2	0.211	1.39	4.61	5.99	9.21	13.82
3	0.584	2.37	6.25	7.81	11.34	16.27
4	1.06	3.36	7.78	9.49	13.28	18.47
5	1.61	4.35	9.24	11.07	15.09	20.52
6	2.20	5.35	10.64	12.59	16.81	22.46
7	2.83	6.35	12.02	14.07	18.48	24.32
8	3.49	7.34	13.36	15.51	20.09	26.13
9	4.17	8.34	14.68	16.92	21.67	27.88
10	4.87	9.34	15.99	18.31	23.21	29.59
11	5.58	10.34	17.28	19.68	24.73	31.26
12	6.30	11.34	18.55	21.03	26.22	32.91
13	7.04	12.34	19.81	22.36	27.69	34.53
14	7.79	13.34	21.06	23.68	29.14	36.12
15	8.55	14.34	22.31	25.00	30.58	37.70
16	9.31	15.34	23.54	26.30	32.00	39.25
17	10.09	16.34	24.77	27.59	33.41	40.79
18	10.86	17.34	25.99	28.87	34.81	42.31
19	11.65	18.34	27.20	30.14	36.19	43.82
20	12.44	19.34	28.41	31.41	37.57	45.32
21	13.24	20.34	29.62	32.67	38.93	46.80
22	14.04	21.34	30.81	33.92	40.29	48.27
23	14.85	22.34	32.01	35.17	41.64	49.73
24	15.66	23.34	33.20	36.42	42.98	51.18
25	16.47	24.34	34.38	37.65	44.31	52.62
26	17.29	25.34	35.56	38.89	45.64	54.05
27	18.11	26.34	36.74	40.11	46.96	55.48
28	18.94	27.34	37.92	41.34	48.28	56.89
29	19.77	28.34	39.09	42.56	49.59	58.30
30	20.60	29.34	40.26	43.77	50.89	59.70
40	29.05	39.34	51.81	55.76	63.69	73.40
50	37.69	49.33	63.17	67.50	76.15	86.66
60	46.46	59.33	74.40	79.08	88.38	99.61
70	55.33	69.33	85.53	90.53	100.43	112.32
80	64.28	79.33	96.58	101.88	112.33	124.84
90	73.29	89.33	107.57	113.15	124.12	137.21
100	82.36	99.33	118.50	123.34	135.81	149.45

Table III The correlation coefficient, r

d.f.	$p = 0.1$	0.05	0.02	0.01	0.005	0.001
1	0.9877	$0.9^{2}692$	$0.9^{3}507$	$0.9^{3}877$	$0.9^{4}692$	$0.9^{5}877$
2	0.9000	0.9500	0.9800	$0.9^{2}000$	$0.9^{2}500$	$0.9^{3}000$
3	0.805	0.878	0.9343	0.9587	0.9740	$0.9^{2}114$
4	0.729	0.811	0.882	0.9172	0.9417	0.9741
5	0.669	0.754	0.833	0.875	0.9056	0.9509
6	0.621	0.707	0.789	0.834	0.870	0.9249
7	0.582	0.666	0.750	0.798	0.836	0.898
8	0.549	0.632	0.715	0.765	0.805	0.872
9	0.521	0.602	0.685	0.735	0.776	0.847
10	0.497	0.576	0.658	0.708	0.750	0.823
11	0.476	0.553	0.634	0.684	0.726	0.801
12	0.457	0.532	0.612	0.661	0.703	0.780
13	0.441	0.514	0.592	0.641	0.683	0.760
14	0.426	0.497	0.574	0.623	0.664	0.742
15	0.412	0.482	0.558	0.606	0.647	0.725
16	0.400	0.468	0.543	0.590	0.631	0.708
17	0.389	0.456	0.529	0.575	0.616	0.693
18	0.378	0.444	0.516	0.561	0.602	0.679
19	0.369	0.433	0.503	0.549	0.589	0.665
20	0.360	0.423	0.492	0.537	0.576	0.652
25	0.323	0.381	0.445	0.487	0.524	0.597
30	0.296	0.349	0.409	0.449	0.484	0.554
35	0.275	0.325	0.381	0.418	0.452	0.519
40	0.257	0.304	0.358	0.393	0.425	0.490
45	0.243	0.288	0.338	0.372	0.403	0.465
50	0.231	0.273	0.322	0.354	0.384	0.443
60	0.211	0.250	0.295	0.325	0.352	0.408
70	0.195	0.232	0.274	0.302	0.327	0.380
80	0.183	0.217	0.257	0.283	0.307	0.357
90	0.173	0.205	0.242	0.267	0.290	0.338
100	0.164	0.195	0.230	0.254	0.276	0.321

Table IV The variance ratio, F $p=0.05$

v_2 \ v_1	1	2	3	4	5	6	7	8	9	10	12	15	20	24	30	40	60	120	∞
1	161.4	199.5	215.7	224.6	230.2	234.0	236.8	238.9	240.5	241.9	243.9	245.9	248.0	249.1	250.1	251.1	252.2	253.3	254.3
2	18.51	19.00	19.16	19.25	19.30	19.33	19.35	19.37	19.38	19.40	19.41	19.43	19.45	19.45	19.46	19.47	19.48	19.49	19.50
3	10.13	9.55	9.28	9.12	9.01	8.94	8.89	8.85	8.81	8.79	8.74	8.70	8.66	8.64	8.62	8.59	8.57	8.55	8.53
4	7.71	6.94	6.59	6.39	6.26	6.16	6.09	6.04	6.00	5.96	5.91	5.86	5.80	5.77	5.75	5.72	5.69	5.66	5.63
5	6.61	5.79	5.41	5.19	5.05	4.95	4.88	4.82	4.77	4.74	4.68	4.62	4.56	4.53	4.50	4.46	4.43	4.40	4.36
6	5.99	5.14	4.76	4.53	4.39	4.28	4.21	4.15	4.10	4.06	4.00	3.94	3.87	3.84	3.81	3.77	7.34	3.70	3.67
7	5.59	4.74	4.35	4.12	3.97	3.87	3.79	3.73	3.68	3.64	3.57	3.51	3.44	3.41	3.38	3.34	3.30	3.27	3.23
8	5.32	4.46	4.07	3.84	3.69	3.58	3.50	3.44	3.39	3.35	3.28	3.22	3.15	3.12	3.08	3.04	3.01	2.97	2.93
9	5.12	4.26	3.86	3.63	3.48	3.37	3.29	3.23	3.18	3.14	3.07	3.01	2.94	2.90	2.86	2.83	2.79	2.75	2.71
10	4.96	4.10	3.71	3.48	3.33	3.22	3.14	3.07	3.02	2.98	2.91	2.85	2.77	2.74	2.70	2.66	2.62	2.58	2.54
11	4.84	3.98	3.59	3.36	3.20	3.09	3.01	2.95	2.90	2.85	2.79	2.72	2.65	2.61	2.57	2.53	2.49	2.45	2.40
12	4.75	3.89	3.49	3.26	3.11	3.00	2.91	2.85	2.80	2.75	2.69	2.62	2.54	2.51	2.47	2.43	2.38	2.34	2.30
13	4.67	3.81	3.41	3.18	3.03	2.92	2.83	2.77	2.71	2.67	2.60	2.53	2.46	2.42	2.38	2.34	2.30	2.25	2.21
14	4.60	3.74	3.34	3.11	2.96	2.85	2.76	2.70	2.65	2.60	2.53	2.46	2.39	2.35	2.31	2.27	2.22	2.18	2.13
15	4.54	3.68	3.29	3.06	2.90	2.79	2.71	2.64	2.59	2.54	2.48	2.40	2.33	2.29	2.25	2.20	2.16	2.11	2.07
16	4.49	3.63	3.24	3.01	2.85	2.74	2.66	2.59	2.54	2.49	2.42	2.35	2.28	2.24	2.19	2.15	2.11	2.06	2.01
17	4.45	3.59	3.20	2.96	2.81	2.70	2.61	2.55	2.49	2.45	2.38	2.31	2.23	2.19	2.15	2.10	2.06	2.01	1.96
18	4.41	3.55	3.16	2.93	2.77	2.66	2.58	2.51	2.46	2.41	2.34	2.27	2.19	2.15	2.11	2.06	2.02	1.97	1.92
19	4.38	3.52	3.13	2.90	2.74	2.63	2.54	2.48	2.42	2.38	2.31	2.23	2.16	2.11	2.07	2.03	1.98	1.93	1.88
20	4.35	3.49	3.10	2.87	2.71	2.60	2.51	2.45	2.39	2.35	2.28	2.20	2.12	2.08	2.04	1.99	1.95	1.90	1.84
21	4.32	3.47	3.07	2.84	2.68	2.57	2.49	2.42	2.37	2.32	2.25	2.18	2.10	2.05	2.01	1.96	1.92	1.87	1.81
22	4.30	3.44	3.05	2.82	2.66	2.55	2.46	2.40	2.34	2.30	2.23	2.15	2.07	2.03	1.98	1.94	1.89	1.84	1.78
23	4.28	3.42	3.03	2.80	2.64	2.53	2.44	2.37	2.32	2.27	2.20	2.13	2.05	2.01	1.96	1.91	1.86	1.81	1.76
24	4.26	3.40	3.01	2.78	2.62	2.51	2.42	2.36	2.30	2.25	2.18	2.11	2.03	1.98	1.94	1.89	1.84	1.79	1.73
25	4.24	3.39	2.99	2.76	2.60	2.49	2.40	2.34	2.28	2.24	2.16	2.09	2.01	1.96	1.92	1.87	1.82	1.77	1.71
26	4.23	3.37	2.98	2.74	2.59	2.47	2.39	2.32	2.27	2.22	2.15	2.07	1.99	1.95	1.90	1.85	1.80	1.75	1.69
27	4.21	3.35	2.96	2.73	2.57	2.46	2.37	2.31	2.25	2.20	2.13	2.06	1.97	1.93	1.88	1.84	1.79	1.73	1.67
28	4.20	3.34	2.95	2.71	2.56	2.45	2.36	2.29	2.24	2.19	2.12	2.04	1.96	1.91	1.87	1.82	1.77	1.71	1.65
29	4.18	3.33	2.93	2.70	2.55	2.43	2.35	2.28	2.22	2.18	2.10	2.03	1.94	1.90	1.85	1.81	1.75	1.70	1.64
30	4.17	3.32	2.92	2.69	2.53	2.42	2.33	2.27	2.21	2.16	2.09	2.01	1.93	1.89	1.84	1.79	1.74	1.68	1.62
40	4.08	3.23	2.84	2.61	2.45	2.34	2.25	2.18	2.12	2.08	2.00	1.92	1.84	1.79	1.74	1.69	1.64	1.58	1.51
60	4.00	3.15	2.76	2.53	2.37	2.25	2.17	2.10	2.04	1.99	1.92	1.84	1.75	1.70	1.65	1.59	1.53	1.47	1.39
120	3.92	3.07	2.68	2.45	2.29	2.17	2.09	2.02	1.96	1.91	1.83	1.75	1.66	1.61	1.55	1.50	1.43	1.35	1.25
∞	3.84	3.00	2.60	2.37	2.21	2.10	2.01	1.94	1.88	1.83	1.75	1.67	1.57	1.52	1.46	1.39	1.32	1.22	1.00

V_1, V_2 are upper, lower d.f. respectively.

Table IV (*cont.*) $p = 0.01$

v_2 \ v_1	1	2	3	4	5	6	7	8	9	10	12	15	20	24	30	40	60	120	∞
1	4052	4999.5	5403	5625	5764	5859	5928	5982	6022	6056	6106	6157	6209	6235	6261	6287	6313	6339	6366
2	98.50	99.00	99.17	99.25	99.30	99.33	99.36	99.37	99.39	99.40	99.42	99.43	99.45	99.46	99.47	99.47	99.48	99.49	99.50
3	34.12	30.82	29.46	28.71	28.24	27.91	27.67	27.49	27.35	27.23	27.05	26.87	26.69	26.60	26.50	26.41	26.32	26.22	26.13
4	21.20	18.00	16.69	15.98	15.52	15.21	14.98	14.80	14.66	14.55	14.37	14.20	14.02	13.93	13.84	13.75	13.65	13.56	13.46
5	16.26	13.27	12.06	11.39	10.97	10.67	10.46	10.29	10.16	10.05	9.89	9.72	9.55	9.47	9.38	9.29	9.20	9.11	9.02
6	13.75	10.92	9.78	9.15	8.75	8.47	8.26	8.10	7.98	7.87	7.72	7.56	7.40	7.31	7.23	7.14	7.06	6.97	6.88
7	12.25	9.55	8.45	7.85	7.46	7.19	6.99	6.84	6.72	6.62	6.47	6.31	6.16	6.07	5.99	5.91	5.82	5.74	5.65
8	11.26	8.65	7.59	7.01	6.63	6.37	6.18	6.03	5.91	5.81	5.67	5.52	5.36	5.28	5.20	5.12	5.03	4.95	5.86
9	10.56	8.02	6.99	6.42	6.06	5.80	5.61	5.47	5.35	5.26	5.11	4.96	4.81	4.73	4.65	4.57	4.48	4.40	4.31
10	10.04	7.56	6.55	5.99	5.64	5.39	5.20	5.06	4.94	4.85	4.71	4.56	4.41	4.33	4.25	4.17	4.08	4.00	3.91
11	9.65	7.21	6.22	5.67	5.32	5.07	4.89	4.74	4.63	4.54	4.40	4.25	4.10	4.02	3.94	3.86	3.78	3.69	3.60
12	9.33	6.93	5.95	5.41	5.06	4.82	4.64	4.50	4.39	4.30	4.16	4.01	3.86	3.78	3.70	3.62	3.54	3.45	3.36
13	9.07	6.70	5.74	5.21	4.86	4.62	4.44	4.30	4.19	4.10	3.96	3.82	3.66	3.59	3.51	3.43	3.34	3.25	3.17
14	8.86	6.51	5.56	5.04	4.69	4.46	4.28	4.14	4.03	3.94	3.80	3.66	3.51	3.43	3.35	3.27	3.18	3.09	3.00
15	8.68	6.36	5.42	4.89	4.56	4.32	4.14	4.00	3.89	3.80	3.67	3.52	3.37	3.29	3.21	3.13	3.05	2.96	2.87
16	8.53	6.23	5.29	4.77	4.44	4.20	4.03	3.89	2.78	3.69	3.55	3.41	3.26	3.18	3.10	3.02	2.93	2.84	2.75
17	8.40	6.11	5.18	4.67	4.34	4.10	3.93	3.79	3.68	3.59	3.46	3.31	3.16	3.08	3.00	2.92	2.83	2.75	2.65
18	8.29	6.01	5.09	4.58	4.25	4.01	3.84	3.71	3.60	3.51	3.37	3.23	3.08	3.00	2.92	2.84	2.75	2.66	2.57
19	8.18	5.93	5.01	4.50	4.17	3.94	3.77	3.63	3.52	3.43	3.30	3.15	3.00	2.92	2.84	2.76	2.67	2.58	2.49
20	8.10	5.85	4.94	4.43	4.10	3.87	3.70	3.56	3.46	3.37	3.23	3.09	2.94	2.86	2.78	2.69	2.61	2.52	2.42
21	8.02	5.78	4.87	4.37	4.04	3.81	3.64	3.51	3.40	3.31	3.17	3.03	2.88	2.80	2.72	2.64	2.55	2.46	2.36
22	7.95	5.72	4.82	4.31	3.99	3.76	3.59	3.45	3.35	3.26	3.12	2.98	2.83	2.75	2.67	2.58	2.50	2.40	2.31
23	7.88	5.66	4.76	4.26	3.94	3.71	3.54	3.41	3.30	3.21	3.07	2.93	2.78	2.70	2.62	2.54	2.45	2.35	2.26
24	7.82	5.61	4.72	4.22	3.90	3.67	3.50	3.36	3.26	3.17	3.03	2.89	2.74	2.66	2.58	2.49	2.40	2.31	2.21
25	7.77	5.57	4.68	4.18	3.85	3.63	3.46	3.32	3.22	3.13	2.99	2.85	2.70	2.62	2.54	2.45	2.36	2.27	2.17
26	7.72	5.53	4.64	4.14	3.82	3.59	3.42	3.29	3.18	3.09	2.96	2.81	2.66	2.58	2.50	2.42	2.33	2.23	2.13
27	7.68	5.49	4.60	4.11	3.78	3.56	3.39	3.26	3.15	3.06	2.93	2.78	2.63	2.55	2.47	2.38	2.29	2.20	2.10
28	7.64	5.45	4.57	4.07	3.75	3.53	3.36	3.23	3.12	3.03	2.90	2.75	2.60	2.52	2.44	2.35	2.26	2.17	2.06
29	7.60	5.42	4.54	4.04	3.73	3.50	3.33	3.20	3.09	3.00	2.87	2.73	2.57	2.49	2.41	2.33	2.23	2.14	2.03
30	7.56	5.39	4.51	4.02	3.70	3.47	3.30	3.17	3.07	2.98	2.84	2.70	2.55	2.47	2.39	2.30	2.21	2.11	2.01
40	7.31	5.18	4.31	3.83	3.51	3.29	3.12	2.99	2.89	2.80	2.66	2.52	2.37	2.29	2.20	2.11	2.02	1.92	1.80
60	7.08	4.98	4.13	3.65	3.34	3.12	2.95	2.82	2.72	2.63	2.50	2.35	2.20	2.12	2.03	1.94	1.84	1.73	1.60
120	6.85	4.79	3.95	3.48	3.17	2.96	2.79	2.66	2.56	2.47	2.34	2.19	2.03	1.95	1.86	1.76	1.66	1.53	1.38
∞	6.63	4.61	3.78	3.32	3.02	2.80	2.64	2.51	2.41	2.32	2.18	2.04	1.88	1.79	1.70	1.59	1.47	1.32	1.00

V_1, V_2, are upper, lower d.f. respectively.

Table V The Studentized Range, Q $p=0.05$

Degrees of Freedom	Number of Treatments																		
	2	**3**	**4**	**5**	**6**	**7**	**8**	**9**	**10**	**11**	**12**	**13**	**14**	**15**	**16**	**17**	**18**	**19**	**20**
1	18.0	27.0	32.8	37.2	40.5	43.1	45.4	47.3	49.1	50.6	51.9	53.2	54.3	55.4	56.3	57.2	58.0	58.8	59.6
2	6.09	8.33	9.80	10.89	11.73	12.43	13.03	13.54	13.99	14.39	14.75	15.08	15.38	15.65	15.91	16.14	16.36	16.57	16.77
3	4.50	5.91	6.83	7.51	8.04	8.47	8.85	9.18	9.46	9.72	9.95	10.16	10.35	10.52	10.69	10.84	10.98	11.12	11.24
4	3.93	5.04	5.76	6.29	6.71	7.06	7.35	7.60	7.83	8.03	8.21	8.37	8.52	8.67	8.80	8.92	9.03	9.14	9.24
5	3.64	4.60	5.22	5.67	6.03	6.33	6.58	6.80	6.99	7.17	7.32	7.47	7.60	7.72	7.83	7.93	8.03	8.12	8.21
6	3.46	4.34	4.90	5.31	5.63	5.89	6.12	6.32	6.49	6.65	6.79	6.92	7.04	7.14	7.24	7.34	7.43	7.51	7.59
7	3.34	4.16	4.68	5.06	5.35	5.59	5.80	5.99	6.15	6.29	6.42	6.54	6.65	6.75	6.84	6.93	7.01	7.08	7.16
8	3.26	4.04	4.53	4.89	5.17	5.40	5.60	5.77	5.92	6.05	6.18	6.29	6.39	6.48	6.57	6.65	6.73	6.80	6.87
9	3.20	3.95	4.42	4.76	5.02	5.24	5.43	5.60	5.74	5.87	5.98	6.09	6.19	6.28	6.36	6.44	6.51	6.58	6.65
10	3.15	3.88	4.33	4.66	4.91	5.12	5.30	5.46	5.60	5.72	5.83	5.93	6.03	6.12	6.20	6.27	6.34	6.41	6.47
11	3.11	3.82	4.26	4.58	4.82	5.03	5.20	5.35	5.49	5.61	5.71	5.81	5.90	5.98	6.06	6.14	6.20	6.27	6.33
12	3.08	3.77	4.20	4.51	4.75	4.95	5.12	5.27	5.40	5.51	5.61	5.71	5.80	5.88	5.95	6.02	6.09	6.15	6.21
13	3.06	3.73	4.15	4.46	4.69	4.88	5.05	5.19	5.32	5.43	5.53	5.63	5.71	5.79	5.86	5.93	6.00	6.06	6.11
14	3.03	3.70	4.11	4.41	4.64	4.83	4.99	5.13	5.25	5.36	5.46	5.56	5.64	5.72	5.79	5.86	5.92	5.98	6.03
15	3.01	3.67	4.08	4.37	4.59	4.78	4.94	5.08	5.20	5.31	5.40	5.49	5.57	5.65	5.72	5.79	5.85	5.91	5.96
16	3.00	3.65	4.05	4.34	4.56	4.74	4.90	5.03	5.15	5.26	5.35	5.44	5.52	5.59	5.66	5.73	5.79	5.84	5.90
17	2.98	3.62	4.02	4.31	4.52	4.70	4.86	4.99	5.11	5.21	5.31	5.39	5.47	5.55	5.61	5.68	5.74	5.79	5.84
18	2.97	3.61	4.00	4.28	4.49	4.67	4.83	4.96	5.07	5.17	5.27	5.35	5.43	5.50	5.57	5.63	5.69	5.74	5.79
19	2.96	3.59	3.98	4.26	4.47	4.64	4.79	4.92	5.04	5.14	5.23	5.32	5.39	5.46	5.53	5.59	5.65	5.70	5.75
20	2.95	3.58	3.96	4.24	4.45	4.62	4.77	4.90	5.01	5.11	5.20	5.28	5.36	5.43	5.50	5.56	5.61	5.66	5.71
24	2.92	3.53	3.90	4.17	4.37	4.54	4.68	4.81	4.92	5.01	5.10	5.18	5.25	5.32	5.38	5.44	5.50	5.55	5.59
30	2.89	3.48	3.84	4.11	4.30	4.46	4.60	4.72	4.83	4.92	5.00	5.08	5.15	5.21	5.27	5.33	5.38	5.43	5.48
40	2.86	3.44	3.79	4.04	4.23	4.39	4.52	4.63	4.74	4.82	4.90	4.98	5.05	5.11	5.17	5.22	5.27	5.32	5.36
60	2.83	3.40	3.74	3.98	4.16	4.31	4.44	4.55	4.65	4.73	4.81	4.88	4.94	5.00	5.06	5.11	5.15	5.20	5.24
120	2.80	3.36	3.69	3.92	4.10	4.24	4.36	4.47	4.56	4.64	4.71	4.78	4.84	4.90	4.95	5.00	5.04	5.09	5.13
∞	2.77	3.32	3.63	3.86	4.03	4.17	4.29	4.39	4.47	4.55	4.62	4.68	4.74	4.80	4.84	4.89	4.93	4.97	5.01

(Reproduced with permission, from Table A15 of SNEDECOR and COCHRANE: *Statistical Methods*, 6th edn. (1967), Iowa State University Press)